T0313179

Volume 2

The Innovation
Tools Handbook

**Evolutionary and Improvement Tools
That Every Innovator Must Know**

Volume 2

The Innovation Tools Handbook

Evolutionary and Improvement Tools That Every Innovator Must Know

Edited by **H. James Harrington** • **Frank Voehl**

CRC Press
Taylor & Francis Group
Boca Raton London New York

CRC Press is an imprint of the
Taylor & Francis Group, an **informa** business

A PRODUCTIVITY PRESS BOOK

CRC Press
Taylor & Francis Group
6000 Broken Sound Parkway NW, Suite 300
Boca Raton, FL 33487-2742

© 2016 by Taylor & Francis Group, LLC
CRC Press is an imprint of Taylor & Francis Group, an Informa business

Printed on acid-free paper
Version Date: 20151203

International Standard Book Number-13: 978-1-4987-6051-5 (Hardback)

Library of Congress Cataloging-in-Publication Data

Names: Harrington, H. J. (H. James), editor. | Voehl, Frank, 1946- editor.
Title: The innovation tools handbook / H. James Harrington and Frank Voehl, editors.
Description: Boca Raton, FL : CRC Press, 2016- | Includes bibliographical references and
  index.
Identifiers: LCCN 2015042020 | ISBN 9781498760492 (vol. 1)
Subjects: LCSH: Technological innovations--Management. | Diffusion of
  innovations--Management. | New products.
Classification: LCC HD45 .I53795 2016 | DDC 658.4/063--dc23
LC record available at http://lccn.loc.gov/2015042020

Visit the Taylor & Francis Web site at
http://www.taylorandfrancis.com

and the CRC Press Web site at
http://www.crcpress.com

I dedicate this book to my son, Jim. As I have grown older I rely more and more on him to keep things running smoothly. In spite of his heavy workload (many nights he is still working at 3 a.m.), he always finds time to call me every evening to be sure I'm okay and wish me good night. Two or three times a week he breaks loose from his normal work schedule to bring me my dinner and spend the evening with me watching television. I thank God that he gave me such a loving and caring son. He really is the sunlight in my life.

H. James Harrington

I dedicate this book to Dr. Myron Tribus, the father of the Quality Council Innovation Movement, who is both a friend and a very powerful influence in my life. He never met a problem he couldn't solve, nor a conundrum he couldn't puzzle out—a deep thinker who taught me that "thinking about thinking" was a goal to be sought after. Thank you for teaching us how to be innovative and to use our creativity for the betterment of man and the world around us.

Frank Voehl

Contents

Foreword

This book is part of a three-book series designed to provide its readers with the tools and methodologies that all innovators should be familiar with and able to use. These are the output from the Tools and Methodologies Working Group of the International Association of Innovative Professionals (IAOIP). The working group was made up of the following individuals:

- H. James Harrington, chairman
- Frank Voehl, co-chairman
- Yared Akalou
- Sifer Aseph
- Scott Benjamin
- Carl Carlson
- Gul Aslan Damci
- Richard Day
- Lisa Friedman
- Thomas Gaskin
- Dallas Goodall
- Luis Guedes
- Paul Hefner
- Dana Landry
- Elena Litovinskaia
- Nikolaos Machairas

- Thomas Mazzone
- Chad McAllister
- Pratik Mehta
- Dimis Michaelides
- Howard Moskowitz
- Michael Phillips
- Jose Carlos Arce Rioboo
- Achmad Rundi
- Robert Sheesley
- Max Singh
- Nithinart Sinthudeacha
- Henryk Stawicki
- Maria Thompson
- Hongbin Wang
- David Wheeler
- Jay van Zyl

The mission statement for the Tools and Methodology Working Group is

Using the expertise and experience of the organization's members and literature research, the working group will define the tools and methodologies that are extensively used in support of the innovation process. The working group will narrow the comprehensive list of tools and methodologies to a list of the ones that are most frequently used in the innovative process and which are the ones that innovative professionals should be confident in using effectively. For each tool and methodology, the working group will prepare a write-up that includes its definition, when it should be used, how to use it, examples of how it has been used, and a list of 5 to 15 questions that can be used to determine if an individual understands the tool or methodology.

To accomplish this mission, the working group studied the literature that was available to define tools and methodologies that were proposed or being used. They also contacted numerous universities that are teaching classes on innovation or entrepreneurship to determine what tools and methodologies they were promoting. In addition, they contacted individual consultants who are providing advice and guidance to organizations in order to identify tools and methodologies they were recommending. As a result of this research, a list of more than 200 tools and methodologies was identified as being potential candidates for the innovative professional.

The group then sent surveys out to leading innovative lecturers, teachers, and consultants, asking them to classify each tool or methodology into one of the following categories:

- This tool or methodology is used on almost all the innovation projects = 4 points.
- This tool or methodology is used on a minimum of two out of five innovation projects = 1 point.
- This tool or methodology is seldom if ever used on innovative product projects = 0 point.
- Not familiar with the tool or methodology = –1 point.
- Never used or recommend this tool or methodology in doing innovation projects = –4 points.

We calculated the priority for each of the tools/methodologies by assigning a point value for each answer. The guidelines that we followed are

- Plus 4 points for a tool/methodology that was always used.
- Plus 1 point for a tool/methodology that is being used at least two out of five projects.
- No points for a tool/methodology that was seldom used.
- Minus 1 for a tool/methodology that the expert had never heard of.
- Minus 4 points for a tool/methodology that the expert never used.

Our goal was to define 50 of the most effective or most frequently used tools/methodologies by the innovative practitioner. We ended up with the 76 tools/methodologies that are the most effective or the most frequently used tools/methodologies by the innovative practitioner (professional).

We then submitted the selected 76 tools/methodologies to a group of 28 practicing innovators, asking them to write a chapter on one or more of the tools/methodologies.

When we assembled the 76 chapters, we ended up with a manuscript of about 1000 pages. After a discussion with the book's editors and key people in the Tools and Methodologies Working Group, it was decided to divide the book up into the following three books:

- Creative tools/methodologies that every innovator should master
- Evolutionary or improvement tools/methodologies that every innovator should master
- Organizational/operational tools/methodologies that every innovator should master

On the basis of these three breakdowns, we went out again to innovative experts asking them to classify each tool as falling into one of the three categories. We soon realized that many of the tools were used in more than one category, so we asked the experts to classify the category that the tool is primarily used in and indicate which categories the tool/methodology could also be used in. Based on this study, we divided the manuscript into three books:

- *Organizational and Operational Tools, Methods, and Techniques That Every Innovator Must Know*
- *Evolutionary and Improvement Tools That Every Innovator Must Know*
- *Creative Tools, Methods, and Techniques That Every Innovator Must Know*

Each book contains the tools/methodologies that were rated as primarily used in that category. The results of this study can be seen in Table F.1.

Genrich Altshuller, the father of TRIZ, did something similar when he analyzed 200,000 patents to determine what unique thought patterns were used to generate the unique patentable idea. On the basis of his study of patents and technological systems, Altshuller proposed that five levels of invention exist:

TABLE F.1

List of the Most Used and/or Most Effective Innovative Tools and Methodologies in Alphabetical Order

Volume 1: Organizational and/or Operational IT&M

Volume 2: Evolutionary and/or Improvement IT&M

Volume 3: Creative IT&M

	IT&M	Volume 3	Volume 2	Volume 1
1.	5 Why questions	S	P	S
2.	76 Standard solutions	P	S	
3.	Absence thinking	P		
4.	Affinity diagram	S	P	S
5.	Agile innovation	S		P
6.	Attribute listing	S	P	
7.	Benchmarking		S	P
8.	Biomimicry	P	S	
9.	Brainwriting 6–3–5	S	P	S
10.	Business case development		S	P
11.	Business plan	S	S	P
12.	Cause-and-effect diagrams		P	S
13.	Combination methods	P	S	
14.	Comparative analysis	S	S	P
15.	Competitive analysis	S	S	P
16.	Competitive shopping		S	P
17.	Concept tree (concept map)	P	S	
18.	Consumer co-creation	P		
19.	Contingency planning		S	P
20.	CO-STAR	S	S	P
21.	Costs analysis	S	S	P
22.	Creative problem solving model	S	P	
23.	Creative thinking	P	S	
24.	Design for tools		P	
	Subtotal—Number of Points	7	7	10

(*Continued*)

TABLE F.1 (CONTINUED)

List of the Most Used and/or Most Effective Innovative Tools and Methodologies in Alphabetical Order

Volume 1: Organizational and/or Operational IT&M

Volume 2: Evolutionary and/or Improvement IT&M

Volume 3: Creative IT&M

	IT&M	Volume 3	Volume 2	Volume 1
25.	Directed/focused/structure innovation	P	S	
26.	Elevator speech	P	S	S
27.	Ethnography	P		
28.	Financial reporting	S	S	P
29.	Flowcharting		P	S
30.	Focus groups	S	S	P
31.	Force field analysis	S	P	
32.	Generic creativity tools	P	S	
33.	HU diagrams	P		
34.	I-TRIZ	P		
35.	Identifying and engaging stakeholders	S	S	P
36.	Imaginary brainstorming	P	S	S
37.	Innovation blueprint	P		S
38.	Innovation master plan	S	S	P
39.	Kano analysis	S	P	S
40.	Knowledge management systems	S	S	P
41.	Lead user analysis	P	S	
42.	Lotus blossom	P	S	
43.	Market research and surveys	S		P
44.	Matrix diagram	P	S	
45.	Mind mapping	P	S	S
46.	Nominal group technique	S	P	
47.	Online innovation platforms	P	S	S
48.	Open innovation	P	S	S
49.	Organizational change management	S	S	P
50.	Outcome-driven innovation	P		
	Subtotal—Number of Points	15	4	7

(*Continued*)

TABLE F.1 (CONTINUED)

List of the Most Used and/or Most Effective Innovative Tools and Methodologies in Alphabetical Order

Volume 1: Organizational and/or Operational IT&M

Volume 2: Evolutionary and/or Improvement IT&M

Volume 3: Creative IT&M

	IT&M	Volume 3	Volume 2	Volume 1
51.	Plan–do–check–act	S	P	
52.	Potential investor present	S		P
53.	Pro-active creativity	P	S	S
54.	Project management	S	S	P
55.	Proof of concepts	P	S	
56.	Quickscore creativity test	P		
57.	Reengineering/redesign		P	
58.	Reverse engineering	S	P	
59.	Robust design	S	P	
60.	S-curve model		S	P
61.	Safeguarding intellectual properties			P
62.	SCAMPER	S	P	
63.	Scenario analysis	P	S	
64.	Simulations	S	P	S
65.	Six Thinking Hats	S	P	S
66.	Social networks	S	P	
67.	Solution analysis diagrams	S	P	
68.	Statistical analysis	S	P	S
69.	Storyboarding	P	S	
70.	Synetics	S	S	P
71.	Systems thinking	P		
72.	Tree diagram	S	P	S
73.	TRIZ	P	S	
74.	Value analysis	S	P	S
75.	Value propositions	S		P
76.	Visioning	S	S	P
	Subtotal—Number of Points	7	12	7
(P) Priority Rating	**Creative**	**Evolutionary**		**Organizational**
Total	29	23		24

IT&M in creativity book: 29

IT&M in evolutionary book: 23

IT&M in organizational book: 24

Note: IT&M, innovative tools and/or methodologies; P, primary usage; S, secondary usage; blank, not used or little used.

- Level 1—Apparent solution
  - Level 1 inventions are obvious and apparent solutions involving well-known methods and knowledge requiring no new invention of any consequence. These are developed based on evolutionary-type thinking patterns.
- Level 2—Minor improvement
  - Level 2 inventions constitute minor nonobvious improvements to a system, using methods known within the domain of discourse but applied in a new way. These are developed based on mostly evolutionary-type thinking patterns.
- Level 3—Major improvement
  - Level 3 inventions involve fundamental improvements to a system involving methods known outside of the domain. This involves applying an idea to the domain that has never been used in the domain previously. These are primarily developed using unique and creative thinking patterns.
- Level 4—New paradigm
  - Level 4 inventions entail the development of an entirely new operating principle and represent radical changes. These are developed using highly creative thinking patterns.
- Level 5—Discovery
  - Level 5 inventions represent a rare scientific discovery or the pioneering of totally new industry altogether. These are developed on the basis of previously untapped concepts and accidental results.

See Figure F.1, which shows the results of Altshuller's analysis.

We were personally surprised to learn that with all of today's focus on creativity and innovation that more than 95% of the patentable ideas are evolutionary in nature and that less than 5% of the patentable ideas are truly creative.

We believe by making effective use of the tools and methodologies presented in this book that an organization can increase the percentage of creative/innovative ideas by five to eight times its present performance level. It is possible; others have done it. It is now up to you to make use of these effective and efficient innovative organizational/operational tools and methodologies.

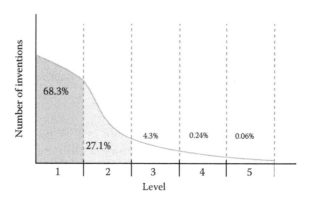

FIGURE F.1
Breakdown of percentage of patents that were developed in each of the five creative classifications.

We are presenting the following 23 tools/methodologies related to innovative evolutionary products, processes, and services, or to improve existing ones.

5 Why questions	Reengineering/redesign
Affinity diagrams	Reverse engineering
Attribute listing	Robust design
Brainwriting 6–3–5	SCAMPER
Cause-and-effect diagrams	Simulations
Creative problem solving model	Six Thinking Hats
Design for tools	Social networks
Flowcharting	Solution analysis diagrams
Force field analysis	Statistical analysis
Kano analysis	Tree diagram
Nominal group technique	Value analysis
Plan–do–check–act	

We are certainly not advocating that you need to use all 23 in order to have an innovative culture. Quite the contrary—the innovator needs to understand all 23, how they are used, and the type or results that they bring about in order to select the right combination for the specific organization they are working with.

H. James Harrington

Preface

In today's fast-moving and high-technology environment, the focus on quality has given way to a focus on innovation. Quality methodology has been shared and integrated into organizations around the world. High quality is now a given for products and services produced in Japan, United States, Germany, Italy, China, India—yes, everywhere. Competition is more fierce and intense than ever before. Technology breakthroughs can be transferred to any part of the world in a matter of days. The people trained in schools around Shanghai are better educated than the people graduating in San Francisco or New York City. The key to being competitive is staying ahead of the competition. That means coming out faster and with more competitive products and services than the competition. The problem that every organization has, be it public or private, profit or nonprofit, product or services, is a need to have more innovative ideas effectively implemented. Today, we need to have our people generate more and better innovative ideas that can be rapidly provided to the consumer. That means that every part of the organization needs to be involved in the innovative activities. Innovative ideas cannot come from just research and development alone. We need to have more innovative processes and systems to support finance, production, sales, marketing, personnel, information technology, procurement—yes, every part of the organization. Even the person sweeping the floors can come up with an innovative idea that will drive a new product cycle. Organizations used to expect their employees, when they came to work, to stop thinking and blindly follow instructions. Today, employees need to realize that they are being paid for both their physical and mental capabilities. Our employees have to understand that if they are going to get ahead, they are now required to be more creative and more innovative at work than any place else. When they get to work, everyone needs to take off their baseball cap and put on their thinking hat in order for the organization to be successful. Today, the best worker is the best thinker, not the one who moves the most products.

Everybody is talking about the importance of innovation, how innovative they are, what innovative products they are producing, and how they need to be more innovative. Everybody is using the word *innovation* to highlight why they are different from everyone else, why customers should rely on them to provide services and product. But what does it all mean? After years of discussion, arguments, and debates, there is little agreement on what a true definition of innovation is. At one extreme, people will argue that innovation is "any new and unique idea." At the other extreme, individuals will define innovation as "a new and unique idea that is produced and delivered to an external customer who is willing to pay more for it than the cost to provide it plus a reasonable profit margin for the supplier." If you use the first definition of innovation, almost all organizations are innovative organizations. If you use the second definition of innovation, less than 5% of the organizations could be considered innovative. Over 95% of all new and unique ideas and suggestions that come from within most organizations never see the light of day; in other words, they never reach the final stage of 'deliverables to the customer.' Accordingly, they never generate a profit for the organization. My preferred definition that may or may not be in keeping with your personal beliefs is, "the process of translating an idea or invention into an intangible product, service, or process that creates value for which the consumer (the entity that uses the output from the idea) is willing to pay more for it than the cost to produce it."

Just so there is no confusion between innovation and creativity, creativity is defined as follows:

- *Creative*—Using the ability to make or think of new things involving the process by which new ideas, stories, products, etc., are created.
- *Create*—Make something; to bring something into existence.

The difference between creativity and innovation is that the output from the innovation has to be a value-added output, while the output from creativity does not have to be value added.

Keeping this information in mind, you can grasp the problem that the International Association of Innovation Professionals (IAOIP) was faced with when they were assigned the responsibility for (a) defining the body of knowledge for innovation and (b) establishing a certification program for innovators. It is obvious that before you can establish a body of knowledge for innovation, you have to have an accepted definition of innovation. Moreover, before you can certify an individual as the innovator, you have to be able to define what an innovator is and does.

Again, the definition of an innovator is very debatable. The following are five different definitions of an innovator:

1. An innovator is an individual who creates a unique idea that is marketable.
2. An innovator is an individual or group who creates a unique idea and is able to guide it through the processes necessary to deliver it to the external customer.
3. An innovator is a person who has the capability to create unique ideas and has the entrepreneurship to turn these ideas into output that is marketed.
4. An innovator is a person who creates a unique idea and uses the facilities available to produce an output that is marketable.
5. An innovator is an individual who creates a unique idea that is marketable and guides it through the development process so that its value to the customer is greater than the resources required to produce it.

As you can see, some of the experts in the field define an innovator as "a person that comes up with a unique and creative idea that adds value." Other experts in the field feel that "an innovator must be capable of finding an unfulfilled need and taking it through each phase of the innovative process." This means that the innovator must be capable of

- Defining an unfulfilled need
- Creating a solution for the unfulfilled need
- Developing a value proposition
- Getting the value proposition approved by management, if it relates to an established organization
- Getting the project funded
- Establishing an organization to produce the output
- Producing the output
- Marketing the output
- Selling the output
- Evaluating the success of the project

Now this may be a lot to expect one person to be able to accomplish. But there are literally millions of these individuals successfully doing this today. You can see some excellent examples of these types of individuals on one of the most-watched new television programs called *Shark Tank*. The contestants are individuals who originated a unique idea, found ways to get it funded, and found ways to produce the output. They also set up the marketing and sales system, and sold the product. These are innovators who have come to the television program to present their innovative output to a group of five entrepreneurs in order to get additional funding and increase sales opportunities. Basically, we believe an innovator is "a person who identifies an unfulfilled need, creates ideas that will fulfill the unmet need, and incorporate the skills of an entrepreneur."

So, what is an entrepreneur? Is an innovator also the same as an entrepreneur? Basically, an entrepreneur is someone who exercises initiative by organizing a venture to take benefit of an opportunity, and, as the decision maker, decides what, how, and how much of a good or service will be produced. An entrepreneur supplies risk capital as a risk taker, and monitors and controls the business activities. The entrepreneur is usually a sole proprietor, a partner, or the one who owns the majority of shares in an incorporated venture. (www.businessdictionary.com)

Keeping this in mind, the real difference between an entrepreneur and an innovator is that the innovator needs to be able to recognize an unfulfilled need and create ideas that fill the unfulfilled need. The entrepreneur does not have to create the idea but can take somebody else's idea and turn it into a value-added output.

As you can see, this discussion applies very nicely to a person who is starting a new organization (e.g., a start-up company), but it is difficult to apply this to an established organization. In an established organization, the innovative process flows through many different functions, and usually the individual who recognized the unfulfilled need and created the idea to fill that unfulfilled need is not assigned to process the idea through the total innovative process. What happens in established companies is that subject matter experts are developed and assigned to individual parts of the innovative process. For example, the controller function is responsible for obtaining adequate financing; marketing defines how the unique concept will be communicated to the customer, and the sales force develops the sales campaign. Manufacturing engineering establishes the production facilities, etc. Each of these functions develops specialized skills to effectively and efficiently process the innovative concept through the innovative cycle. It is a rare exception within an established company when the individual who created the innovative idea is held responsible for having it progress through the innovative cycle and create value added to the consumer and the organization. In the cases where this occurs, that individual is called an entrepreneur. (An entrepreneur is an employee of a large corporation who is given freedom and financial support to create new products, service, systems, etc., and does not have to follow the corporation's usual routines or protocols.) In this book, we will be using the following definition of an innovator:

> An innovator is an individual who creates a unique idea that is marketable; one who then guides it through the innovative process so that its value to the customer is greater than the resources required to produce it.

Considering all the difficulties there are in getting an agreed-to definition of innovation or innovator, you might think it would be impossible to define what types of tools and methodologies should be used during the innovative cycle. Not so—we accomplished this by forming a committee to research key documentation related to books and technical articles on innovation methods and techniques. We then prepared a list of recommended tools and methodologies and distributed the list to many of the experts who were teaching or using the innovation process. Each of these experts, in turn, evaluated the long list of tools and methodologies that we had collected, and each put these tools in the following categories:

- Always used
- Frequently used
- Seldom used
- Never used
- Not known

We also asked them to suggest any additional tool/methodology that was missing from the list. As a result of this research study, we defined the tools that were most efficient, effective, and frequently used in the innovative process. These tools are presented in this book along with enough information to show you how to use them. Each tool is represented by a chapter and presented in the following format:

- Definition: tool and/or methodology.
- User: who uses the tool/methodology.
- What phases of the innovative process are the tool/methodology used in?

- How is the tool/methodology used?
- Examples of the outputs from the tool/methodology.
- Software to help in using the tool/methodology.
- References.
- Suggested additional reading.

For many of the tools, there are, or should be, complete books written on how to effectively utilize the tool/methodology. For example, there are a number of books written on storyboarding. In most cases, we have recommended additional reading for those who desire more detailed information on how to effectively implement and use the tool/methodology. We do not believe that most people involved in the innovative process will need to master all of the tools or methodologies listed in this book; however, we do believe that all of them are important enough that all the individuals involved in the innovative process should at least be familiar with each of them.

We also recommend that anyone actually involved in moving the innovative idea through any part of the innovative process become active members in the IAOIP. Their mission is to organize and advance innovation through the development of a catalog of innovation skills and capabilities as well as certifications to demonstrate mastery of that body of knowledge. Working groups are organized to create the base of advanced innovation certification requirements and ensure our certified members and organizations are global leaders in practicing and managing innovation. More information related to the IAOIP can be found by going to http://iaoip.org/.

H. James Harrington

Acknowledgments

We acknowledge the many hours of work that each of the innovative professionals who wrote chapters for this book expended without compensation—done in order to capture and share the knowledge contained in this book. We would be remiss not to acknowledge the many long days and late evenings that Candy Rogers, Joe Mueller, and Susan Koepp-Baker put into proofreading and formatting this book. It was a major challenge to convert the creative thinking of so many individuals into a standard pattern so that the book flowed freely and logically from chapter to chapter.

This book represents the output from the Tools and Methodologies Working Group of the International Association of Innovative Professionals (IAOIP).

About the Editors

 **Dr. H. James Harrington**, chief executive officer (CEO), Harrington Management Systems. In the book *Tech Trending*, Dr. Harrington was referred to as "the quintessential tech trender." The *New York Times* referred to him as having a "...knack for synthesis and an open mind about packaging his knowledge and experience in new ways—characteristics that may matter more as prerequisites for new-economy success than technical wizardry...."

It has been said about him, "Harrington writes the books that other consultants use."

The leading Japanese author on quality, Professor Yoshio Kondo, stated, "Business Process Improvement (methodology) investigated and established by Dr. H. James Harrington and his group bring some of the new strategies which brings revolutionary improvement not only in quality of products and services, but also the business processes which yield the excellent quality of the output."

The father of *total quality control*, Dr. Armand V. Feigenbaum, stated, "Harrington is one of those very rare business leaders who combines outstanding inherent ability, effective management skills, broad technology background and great effectiveness in producing results. His record of accomplishment is a very long, broad and deep one that is highly and favorably recognized."

Bill Clinton, as president of the United States, appointed Dr. Harrington to serve as an Ambassador of Goodwill.

Newt Gingrich, former Speaker of the House and general chairman of American Solutions, has appointed Dr. H. James Harrington to the advisory board of his Jobs and Prosperity Task Force.

KEY RESPONSIBILITIES

H. James Harrington now serves as the CEO for Harrington Management Systems, and he is on the board of directors for a number of small- to medium-size companies helping them develop their business strategies. He also serves as

- President of the Walter L. Hurd Foundation
- Honorary advisor for quality for China
- Chairman of the Centre for Organizational Excellence Research (COER)
- President of the Altshuller Institute

AWARDS AND RECOGNITION

Harrington received many awards and recognition trophies throughout his 60 years' activity in promoting quality and high performance throughout the world. He has had many performance improvement awards named after him from countries worldwide. Some of them are as follows:

- The Harrington/Ishikawa Medal, presented yearly by the Asian Pacific Quality Organization, was named after H. James Harrington to recognize his many contributions to the region.
- The Harrington/Neron Medal was named after H. James Harrington in 1997 for his many contributions to the quality movement in Canada.
- Harrington Best TQM Thesis Award was established in 2004 and named after H. James Harrington by the European Universities Network and e-TQM College.
- Harrington Chair in Performance Excellence was established in 2005 at the Sudan University.
- Harrington Excellence Medal was established in 2007 to recognize an individual who uses the quality tools in a superior manner.
- H. James Harrington Scholarship was established in 2011 by the ASQ Inspection Division.

PUBLICATIONS AND LECTURES

Harrington is the author of more than 40 books and hundreds of papers on performance improvement of which more than 150 have been published in major magazines. He has given hundreds of seminars on every continent of the South Pole.

Frank Voehl, president, Strategy Associates, now serves as the chairman and president of Strategy Associates Inc. and as a senior consultant and chancellor for Harrington Management Systems. He also serves as the chairman of the board for a number of businesses and as a Grand Master Black Belt instructor and technology advisor at the University of Central Florida in Orlando, Florida. He is recognized as one of the world leaders in applying quality measurement and Lean Six Sigma methodologies to business processes.

PREVIOUS EXPERIENCE

Frank Voehl has extensive knowledge of National Regulatory Commission, Food and Drug Administration, Good Manufacturing Practice, and National Aeronautics and Space Administration quality system requirements. He is an expert in ISO-9000, QS-9000/14000/18000, and integrated Lean Six Sigma quality system standards and processes. He has degrees from St. John's University and advanced studies at New York University (NYU), as well as an honorary doctor of divinity degree. Since 1986, he has been responsible for overseeing the implementation of quality management systems with organizations in such diverse industries as telecommunications and utilities, federal, state and local government agencies, public administration and safety, pharmaceuticals, insurance/banking, manufacturing, and institutes of higher learning. In 2002, he joined The Harrington Group as the chief operating officer (COO) and executive vice president. He has held executive management positions with Florida Power and Light and FPL Group, where he was the founding general manager and COO of QualTec Quality Services for 7 years. He has written and published/co-published more than 35 books and hundreds of technical papers on business management, quality improvement, change management, knowledge management, logistics and teambuilding, and has received numerous awards for community leadership, service to third-world countries, and student mentoring.

CREDENTIALS

The Bahamas National Quality Award was developed in 1991 by Voehl to recognize the many contributions of companies in the Caribbean region, and he is an honorary member of its board of judges. In 1980, the City of Yonkers, New York, declared March 7 as *Frank Voehl Day*, honoring him for his many contributions on behalf of thousands of youth in the city where he lived, performed volunteer work, and served as athletic director and coach of the Yonkers-Pelton Basketball Association. In 1985, he was named *Father of the Year* in Broward County, Florida. He also serves as president of the Miami Archdiocesan Council of the St. Vincent de Paul Society, whose mission is to serve the poor and needy throughout South Florida and the world.

About the Contributors

Stuart Burge is one of the founding partners of the consultancy and training company Burge Hughes Walsh (BHW). He is widely known in industry and academia for his expertise in systems engineering and particularly systems design. Stuart is recognized for his pragmatic approach to systems engineering and his ability to explain how to actually do practical systems engineering. Since the formation of BHW in 2000, Stuart has worked with a large number of clients, helping them improve their systems engineering. This has been through the design and delivery of training courses, coaching and facilitating individuals and project teams, and undertaking research into systems engineering and systems design. Stuart is also a Six Sigma Master Black Belt with particular expertise in Design for Six Sigma and robust design.

Neil Farmer is former head of Research Reports at Butler Cox and founder of Farringdon Research. He joined Informal Networks as change implementation and research director in 2009. Neil is a very unusual hybrid of researcher and practical implementer. He has been responsible for change management at some of the most successful business and organizational transformations during the last 15 years. Neil is the author of two books: *Total Business Design*, published by Wiley in 1996, and *The Invisible Organization*, published by Gower in 2008.

Dimis Michaelides is the managing director of Performa Consulting. After 22 years at the World Bank in Washington D.C., at ICI (now Syngenta—a chemical multinational) in Paris, and in the position of CEO at Laiki Cyprialife (a life insurance company) in Nicosia, Dimis created Performa, a consulting and training company whose work is centered around creativity and innovation. Dimis has been a visiting professor at a number of business schools and universities (including

Queen Mary and Royal Holloway—University of London, CKGSB—Beijing, JiaoTong—Shanghai, and INSEAD—Fontainebleau) and has served on various boards of directors (including the board of trustees of The Creative Education Foundation). Dimis is a frequent keynote speaker in conferences around the world.

Charles Mignosa is vice president of management solutions in Harrington Management Systems. Mr. Mignosa has more than 30 years of diversified experience in high technology, telecommunications, food processing, and biomedical device industries, and 25 years of experience in IBM holding patents in solid lubricants. In addition to a BS in chemistry, Mr. Mignosa has graduate degrees in statistics and systems research and is a senior member of the American Society for Quality (ASQ). He is a Master Six Sigma Black Belt.

Douglas Nelson is an experienced consultant and manager with international experience in strategic planning, business development, business problem solving, and providing results-oriented management solutions. He is a Lean Six Sigma Master Black Belt/Process Excellence Professional with proven ability and experience to drive operational process improvement. He has more than 25 years of management and operational experience providing process, cost, and productivity improvement in the manufacturing and service industries.

Achmad Rundi is the director of Technology and Innovation at Catalyst Global Consulting. He earned his master's degree in management of technology from NYU Polytechnic School of Engineering.

Peter Westbrook is an experienced manage-ment consultant, mentor, and facilitator, primarily focused on strategic change, quality of products and services, and improving business outcomes. He has particular expertise in strategic thinking, collabora-tive partnerships, and mapping and measuring the effectiveness of an organization's informal relation-ships (social capital) and formal relationships, busi-ness self-assessment, leadership development, system thinking, quality assurance, improvement techniques, and facilitating development and problem-solving workshops. Peter is also well connected to a wide variety of people who have leading edge skills or management tools. He draws on his experience, expertise, and networks to work effectively with clients to create the most effective solution. Consequently, he has been instrumen-tal in delivering significant business change resulting in more effective leadership, strategic plans, cost savings, improved project delivery times, better quality, and greater customer satisfaction.

1

5 Whys

Frank Voehl

CONTENTS

DEFINITION

The 5 Whys is a technique to get to the root cause of the problem. It is the practice of asking five times or more why the failure has occurred in order to get to the root cause. Each time an answer is given, you ask why that particular condition occurred. As outlined in this chapter, it is recommended that the 5 Whys be used with risk assessment in order to strengthen the use of the tool for innovation and creativity-enhancing purposes.

USER

This tool can be used by individuals, but its best use is with a group of four to eight people. Cross-functional teams usually yield the best results from this activity.

OFTEN USED IN THE FOLLOWING PHASES OF THE INNOVATIVE PROCESS

The following are the seven phases of the innovative cycle. An X after the phase name indicates that the tool/methodology is used during that specific phase.

- Creation phase X
- Value proposition phase
- Financing phase
- Documentation phase
- Production phase X
- Sales/delivery phase X
- Performance analysis phase X

TOOL ACTIVITY BY PHASE

- Creation phase—During this phase, the 5 Whys technique is used to get to the root cause of questions like, "Why does the customer want that?"
- Production phase—During this phase, the 5 Whys is used to get to the root cause of questions related to problems that are defined and implementing the proposed initiative.
- Sales/delivery phase—During this phase, the 5 Whys tool is used to get to the root cause of problems like, "Why aren't the customers buying this product?"
- Performance analysis phase—During this phase, the tool is used to answer questions like, "Why was the return on investment so low?" or "Why was this product so profitable?"

HOW TO USE THE TOOL

As previously mentioned, remember that the ease of using the 5 Whys method and the time required for its processing have to be balanced with the potential for the problem to recur if the 5 Whys does not work. In other words, one of the problems with using the 5 Whys is that it does not always uncover the root causes when the cause(s) is not known. This is why some type of risk assessment should always be used with the 5 Whys.

You should also remember that the 5 Whys method presumes that each symptom has one cause, and yet in many cases, this is not true. The result is that this type of analysis does not always show several related variables that are causing the symptom. Also, how well the 5 Whys works for innovation depends, to a degree, on the skill of the person using it. If one element of the 5 Whys has an incorrect answer, it can throw off the entire analysis. Lastly, this method is very often not easily repeatable. For example, having three people applying the 5 Whys can often come up with two or three different answers.

Three possible ways to strengthen the 5 Whys include

- Gather data and evidence to demonstrate why the answer to any of the 5 Whys is likely and plausible versus not likely.
- Couple data analysis with good risk assessment.
- Come up with a baseline with time frames of the particular events that detail how the problem happened or the opportunity unfolded.

When you gather evidence to support an answer to any of the 5 Whys, you are avoiding the trap of falling into a deductive thinking and reasoning mode to answer one of the 5 Whys. You should also add a test loop to every one of the Why levels that attempts to validate the answer through evidence. A risk assessment with your 5 Whys tool allows you to cut down on the weakness in the method in those cases where the true root cause is not found. Also, if the 5 Whys technique does produce a likely root cause, yet there is uncertainty regarding the true root cause, a risk assessment can be useful to see if a more stringent root cause analysis should be pursued in order to do a more stringent analysis.

The 5 Whys was originally adopted from Japanese management systems as a problem-solving tool and can be useful for solving hidden problems that in many cases prove to be symptoms of underlying hidden issues that often escape notice. While a quick fix is often adopted, in many problem situations, it may well be a poor, not very useful solution, and in the final analysis may not solve any or just part of the problem. To solve the problem with a complete solid outcome, you need to systematically do *single-case boring* down through the layers upon layers of symptoms to eventually reach the underlying cause. As originally developed by Sakichi Toyoda, the 5 Whys technique is a straightforward and powerful tool for systematically uncovering the root of a problem so that it can be dealt with effectively and prevent it from recurring over and over again.

As previously mentioned, Sakichi Toyoda was the first to develop the 5 Whys tool in the 1930s. As the founder of Toyota Industries, he was a respected businessman whose technique eventually became popular in the 1970s; Toyota, along with most of the world, still uses it to solve problems today. Toyota has a *go and see* philosophy. This means that its decision making is based on an in-depth understanding of the processes and conditions on the shop floor, rather than reflecting what someone in a boardroom thinks might be happening.

The 5 Whys technique is true to this tradition, and it is most effective when the answers come from people who have hands-on experience of the process being examined. It is remarkably simple: when a problem occurs, you uncover its nature and source by asking *why* no fewer than five times. The 5 Whys is a simple, practical tool that is very easy to use. When a problem arises, the suggestion is to keep asking the question *why* until you reach the underlying source of the problem, which sometimes requires more than one path to reach the root cause(s). In some cases, the term *countermeasure* is used to designate an action or combination of actions that seeks to prevent the problem from surfacing once again, while a solution just seeks to deal with the immediate problem situation. As such, the use of countermeasures is considered by many to be more robust, and more likely to prevent the problem from recurring.

Both individuals and groups can use the 5 Whys in the innovation/idea creation process (is this a good idea, and why?), and in troubleshooting (is this a significant or important issue?), quality improvement (are there quality issues that are surfaced?), and also in problem solving (is the root cause of the problem observable or detectable?). Many people find that it is best for simplistic or not-too-difficult opportunities. (Note: In many cases involving more complex or critical problems, the 5 Whys can lead you to pursue a single track of inquiry when there could be multiple causes. This is overcome by pursuing multiple paths, as the Case Study at the end of this chapter indicates.)

The simplicity of the 5 Whys tool gives it great flexibility, too, and it combines well with other methods and techniques; it is often a first choice before embarking on more complex tool usage. Each time you ask *why*, you need to look for an answer that is grounded in data and fact, as it should be an accounting of things that have actually happened and not speculation on events that might have happened. This approach focuses the 5 Whys on becoming just a process of deductive reasoning that can generate a number of possible causes and can sometimes create more confusion. Keep asking *why* until you feel confident that you have identified the root cause and can go no further. At this point, an appropriate countermeasure should become evident.

Advantages

- The 5 Whys technique is applied with relative ease, making it a simple-to-use practical tool for root cause analysis in problem solving.
- With frequent practice, it is possible to get to root causes in a relatively short period of time.
- Unlike more sophisticated innovation techniques, the 5 Whys methodology does not involve advanced statistical tools or sophisticated data segmentation.
- By repeatedly asking *why* four or five times, the essence of the problem and its associated solution become obvious.
- By repeatedly asking *why* multiple times, you can peel away in a systematic manner the various layers of symptoms, which then can lead you to identify the opportunities you are looking for, along with the root causes of a problem.

Disadvantages

- While many companies have successfully used this tool, the method has some inherent limitations, which include the inability to distinguish between causal factors and the root cause, and a lack of rigor where the user is not mandated to do sufficiency testing.
- Using 5 Whys does not always lead to root cause identification when the cause or the opportunity is unknown.

- If the cause is unknown to the person or group doing the innovation, using 5 Whys may not lead to any meaningful answers.
- An assumption underlying 5 Whys is that each presenting symptom has only one sufficient cause. This is not always the case, and the 5 Whys analysis may not reveal jointly sufficient causes that explain a symptom.
- The success of 5 Whys is, to some degree, contingent upon the skill with which the method is applied; if even one *why* has a bad or meaningless answer, the whole procedure can be thrown off.
- The method is not necessarily repeatable; three different people applying 5 Whys to the same problem may come up with three totally different answers.
- Other drawbacks to 5 Whys have been cited, including the method's inability to distinguish between causal factors and root causes, and the lack of rigor where users are not required to test for sufficiency the root causes generated by the method.

EXAMPLE

See Figure 1.1.

CASE STUDY

National Aeronautics and Space Administration O-Ring Failure

If you ask anyone familiar with the details of the 1986 Space Shuttle Challenger disaster what caused the shuttle accident, they will likely say "the O-rings." While the *O-rings* is a necessary part of the explanation, it is insufficient in providing a complete picture of what led up to that fateful moment. You may say "that's all I want to know," and that is fine; there is nothing that says that all of the details must be understood by everyone. But for those who want to know the complete story, the 5 Whys technique can help.

The National Aeronautics and Space Administration (NASA) has provided the background information and the facts used to develop this

Questions		Write your answers here
What is the problem statement?		The firefighting process did not meet the Muda City citizen's requirements of putting out the fire in a timely manner, with minimal damage to the building structure, resulting in a great deal of wasted time and total loss of the building, along with poor documentation of the event. This caused the fire chief to become upset with the crew leader as the needed paperwork was handled in a sloppy and unprofessional manner.
Why did the problem happen?	A	It took too long for the firefighters to reach the proper destination and put out the fire.
Why did A happen?	B	The fire was not reported in a timely manner, and was not routed to the right fire station.
Why did B happen?	C	The dispatch system was not effective, causing delays in assigning the responsibility and reaching and putting out the fire.
Why did C happen?	D	The dispatch system was not integrated, and knowledge of the key impacts was not known or made available in a timely manner.
Why did D happen?	E	The Muda City's knowledge management system was not automated or up-to-date.
Why did E happen?		The city did not have adequate money in the current budget to provide for an affective firefighting knowledge management system that is integrated and aligned with the people, processes, and systems required.

FIGURE 1.1
5 Whys example.

example.* The following is the 5 Whys analysis tree based on the NASA information.

1. Why did the Space Shuttle Challenger explode? Because the external hydrogen tank ignited owing to the hot gases leaking from one of the solid booster rocket motors.
2. Why did hot gases leak from one of the solid booster rocket motors? Because the seal between the two lower segments of the rocket motor failed to prevent the leak.
3. Why did the seal fail? Because the O-ring, which was intended to compensate for variations in the seal between the segments, failed owing to effects of extreme temperature, which was below 30°F, and was never tested for this low temperature.
4. A: Why did the O-ring fail its intended purpose? Because of a known design flaw with the seal.
5. B: Why was it not tested for lower than 30°F? Because the weather in Cape Canaveral, Florida, never went below 30°F at that time of the year.
6. Why did the mission proceed with known and unknown design flaws? The pressure to launch became too great, and there were serious flaws in the launch decision-making process, as well as shortcuts being taken in the components-testing areas. The documented ones were failure to adequately address problems that require corrective action; both NASA and Thiokol accepted escalating risk apparently because they "got away with it last time."†

Evidence and Risk Assessment

1. Note that the structure of the analysis tree in our NASA case study consisted of a *why* question followed by a theory that answers that question. On the basis of that theory, a new *why* question was formulated and a theory about that question is clearly stated.
2. This process continues for as long as the question remains relevant. It is important that the theory be stated in terms that would enable

---

* Report of the Presidential Commission on the Space Shuttle Challenger Accident. Information is in the public domain and can be freely used.
† Prof. Richard Fenyman wrote about the lack of effective management oversight in his Appendix to the Challenger Accident Report.

a *why* question to follow. For example, if the question asked earlier about what caused the accident, and the answer "the O-rings" were left without a challenge being issued, we would be left without any additional insights.

3. The challenge should be, "What about the O-rings?" and the response might be, "Well, they failed." Now we can ask the question, "Why did they fail?" This is where the process usually breaks down; we sometimes allow our impatience to outweigh our curiosity, and do not want to annoy anyone with obvious and obnoxious questions.

4. But, as you can see in the analysis, persistence revealed the fact that this was a known problem. The reality was that the mission was allowed to proceed anyway.

5. The explanation in the analysis actually provides a more comprehensive explanation for the O-ring failure, because we also challenged the upstream premises. For example, the question should also be asked:

   "How could an O-ring failure cause the shuttle to explode?"

   "This caused a seal failure in the segments of one of the solid rocket motors."

   "Go on."

   "That caused hot gases to escape onto the external hydrogen tank."

   Now we have a much richer explanation for the whole episode.

6. Note that while the *why* question is used to traverse the tree in a downward direction to get to the ultimate root cause, we can use *because* to help explain the conclusion from the bottom up. For example, because there were serious flaws in the launch decision-making process, the mission was allowed to proceed with a known design flaw.

7. With this more comprehensive picture of what really happened, the person who was investigating the O-ring that caused the failure now may want to know more about the decision-making process at NASA, and a new set of 5 Whys begins in a circulatory pattern.

SOFTWARE

No software is required.

SUGGESTED ADDITIONAL READING

Asaka, T. and Ozeki, K., eds. *Handbook of Quality Tools: The Japanese Approach.* Portland, OR: Productivity Press, 1998.

Harrington, H.J. and Lomax, K. *Performance Improvement Methods.* New York: McGraw-Hill, 2000.

2

Affinity Diagrams

H. James Harrington

CONTENTS

Affinity diagram—very complicated but in reality quite simple to master.

H. James Harrington

DEFINITION

Affinity diagrams are a technique for organizing a variety of subjective data into categories based on the intuitive relationships among individual pieces of information. It is often used to find commonalties among concerns and ideas. It lets new patterns and relationships between ideas be discovered (Harrington and Lomax, 2000).

USER

This tool can be used by individuals, but its best use is with a group of four to eight people. Cross-functional teams usually yield the best results from this activity.

OFTEN USED IN THE FOLLOWING PHASES OF THE INNOVATIVE PROCESS

The following are the seven phases of the innovative cycle. An X after the phase name indicates that the tool/methodology is used during that specific phase.

- Creation phase X
- Value proposition phase X
- Resourcing phase X
- Documentation phase
- Production phase X
- Sales/delivery phase
- Performance analysis phase

TOOL ACTIVITY BY PHASE

- Creation phase—During this phase, this tool can be used to look at different options, comparing their advantages and disadvantages to select the best option.
- Value proposition phase—During this phase, this tool can be used to look at and evaluate the interrelationship between potential outputs and outcomes.
- Resourcing phase—During this phase, this tool can be used to compare different financing options so that the least risk and minimum use of assets are used related to obtaining the required funding.
- Production phase—During this phase, this tool can be used to optimize the decision-making process related to the many options for

producing the output. For example, should it be produced internally or should it be outsourced to Asia?

HOW TO USE THE TOOL

Affinity diagrams are used to organize the large numbers of ideas derived from brainstorming. While man has been combining ideas into relational groups for years, it was not until 1960 when Jiro Kawakita formalized the activity, coining the term *affinity diagram*. The process is sometimes referred to as the *KJ method* (Harrington and Lomax, 2000).

Start by defining the issue and gathering random ideas from a brainstorming session.

- Identify areas of concern.
- Organize the ideas into the concern categories.
- Prioritize the ideas in each category.
- Form subgroups as ideas are sorted.

The completed affinity diagram can be used as input to a cause-and-effect diagram.

The following are some of the typical times when affinity diagrams provide a useful tool.

- When you are confronted with many facts or ideas in apparent chaos
- When analyzing survey results
- When analyzing from a brainstorming exercise
- When a lot of complex data are available and need to be organized

PROCESS TO PREPARE AN AFFINITY DIAGRAM

- Start by ensuring that the materials are available to do an affinity diagram. You will need marking pens, a large working surface like a wall or a large table, and a good supply of white cards or posted notes.
- Have each individual on the team record each of his or her ideas on separate cards or posted notes.

- Have each member of the team lay out his or her recorded ideas in a random matter on a working surface.
- Have the team look for ideas that seem to be related in some way. Place them side by side. Repeat until all notes are grouped.
- Have the team discuss the layout of the chart, especially the reasons for moving notes into their specific groupings. This usually results in more changes to the diagram.
- When ideas are grouped, select a heading for each group. Often, it is necessary to make up a special card that summarizes the thought pattern for the entire grouping. It is useful to identify this name card for the grouping by putting it on a different color piece of paper or highlighting it in some manner.
- The team should now look at the title cards to see if some of them should be combined to minimize the number of issues being discussed.

EXAMPLE

- Start by defining the issue and gathering random ideas from a brainstorming session (see Figure 2.1).
- Identify areas of concern (see Figure 2.2).
- Form subgroups as ideas are sorted and organize the ideas into the concern categories (see Figure 2.3).
- Separate the concern categories into actionable categories (see Figure 2.4).
- Identify actionable areas (see Figure 2.5).

SOFTWARE

Some commercial software available includes but is not limited to

- Edraw max: http://www.edrawsoft.com
- Smartdraw: http://www.smartdraw.com
- Affinity Diagram 2.1: http://mobile.brothersoft.com/
- QI macros: http://www.qimacros.com

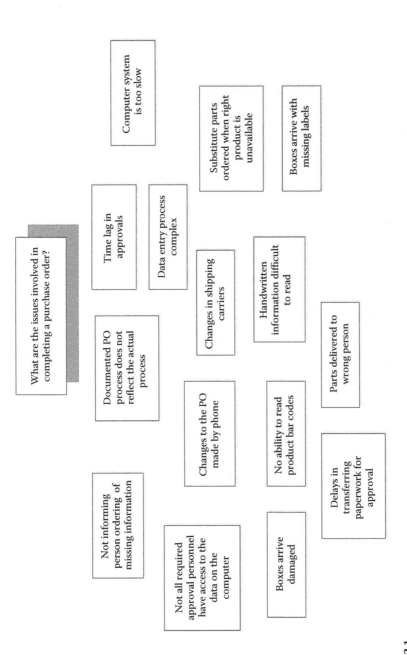

FIGURE 2.1
Brainstorming session idea generation.

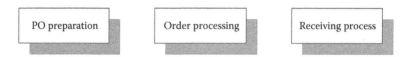

FIGURE 2.2
Areas of concern identified.

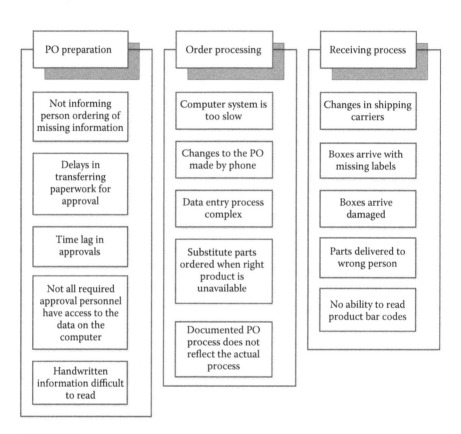

FIGURE 2.3
Concern categories.

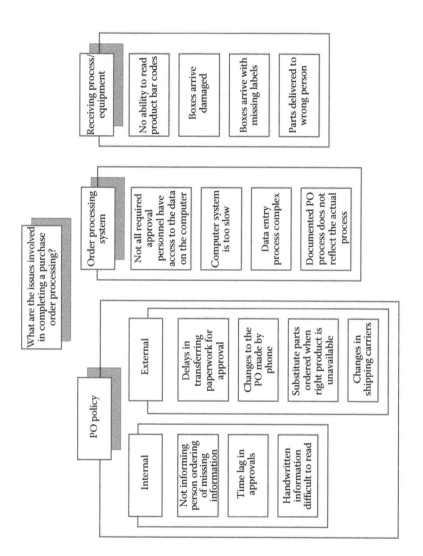

FIGURE 2.4

Concern categories separated into actionable categories.

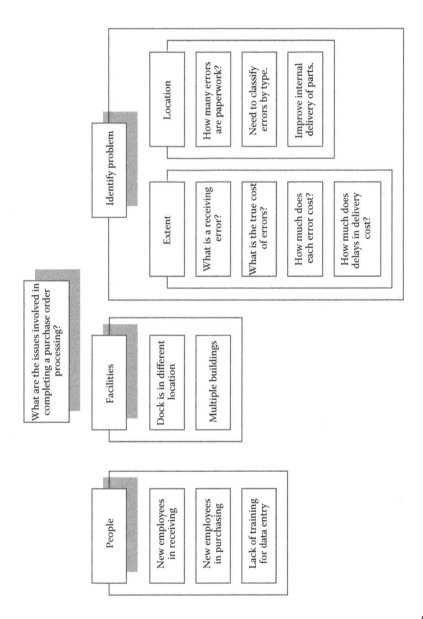

FIGURE 2.5
Actionable categories identified.

REFERENCE

Harrington, H.J. and Lomax, K. *Performance Improvement Methods*. New York: McGraw-Hill, 2000.

SUGGESTED ADDITIONAL READING

Asaka, T. and Ozeki, K., eds. *Handbook of Quality Tools: The Japanese Approach*. Portland, OR: Productivity Press, 1998.
Brassard, M. *The Memory Jogger Plus*. Milwaukee, WI: ASQ Quality Press, 1989.
Eiga, T., Futami, R. Miyawama, H., and Nayatani, Y. *The Seven New QC Tools: Practical Applications for Managers*. New York: Quality Resources, 1994.
King, B. *The Seven Management Tools*. Methuen, MA: Goal/QPC, 1989.
Mizuno, S., ed. *Management for Quality Improvement: The 7 New QC Tools*. Portland, OR: Productivity Press, 1988.

3

Attribute Listing, Morphological Analysis, and Matrix Analysis

Frank Voehl

CONTENTS

DEFINITION

Attribute listing, morphological analysis, and matrix analysis techniques are good for finding new combinations of products or services. We use attribute listing and morphological analysis to generate new products and services. Matrix analysis focuses on businesses. It is used to generate new approaches, using attributes such as market sectors, customer needs, products, promotional methods, etc.

USER

This tool can be used by individuals or groups, but its best use is with a group of four to eight people. Cross-functional teams usually yield the best results from this activity. Since attribute listing is a creative method to find new ideas to solve problems and to find innovative products and services, this method can be very usefully combined with a brainstorming session. The core idea is for a team to forget everything they have learned and retained when thinking about a possible solution, that is, wipe the slate clean.

OFTEN USED IN THE FOLLOWING PHASES OF THE INNOVATIVE PROCESS

The following are the seven phases of the innovative cycle. An X after the phase name indicates that the tool/methodology is used during that specific phase.

- Creation phase X
- Value proposition phase X
- Resourcing phase
- Documentation phase
- Production phase X
- Sales/delivery phase
- Performance analysis phase

TOOL ACTIVITY BY PHASE

- Creation phase—During this phase, this tool is used to generate ideas that will lead to new and unforeseen products and services.
- Value proposition phase—During this phase, this tool is used to define potential markets and marketing approaches in order to improve the accuracy of potential revenue and risks related to the product or service.

- Production phase—During this phase, this tool is primary used in the problem-solving mode.
- Sales/delivery phase—During this phase, this tool is used for defining marketing and sales approaches and to help define market segments.

HOW TO USE THE TOOL

Attribute listing is a creative technique used to find new ideas, solve problems, and find innovative products and services. It usually involves breaking the problem down into smaller parts and looking at alternative solutions. It is often used in the research & development (R&D) departments of many companies, especially those who are constantly looking to produce innovative products to have an advantage over their competitors. It is one of the best ways to generate ideas, especially if there are many parts to the problem or challenge faced. The usual procedure is to take an existing product or service and list all of its attributes.

Attribute listing and morphological analysis are good techniques for finding new combinations of products or services. They are sufficiently similar to be discussed together. We use attribute listing and morphological analysis to generate new products and services. It is a creative technique geared to find new ideas, to solve problems, and to find innovative products and services. This technique is very often combined with other knowledge management tools like brainstorming and brainwriting sessions. It can also be combined with tools to explore knowledge sources and tools to identify existing expert knowledge in the company so as to support the successful composition of working groups.

To use these tools, first list the attributes of the product, service, or strategy you are examining. Attributes are parts, properties, qualities, or design elements of the thing being looked at. For example, attributes of a pencil would be shaft material, lead material, hardness of lead, width of lead, quality, color, weight, price, and so on, as shown in the provided Examples. A television plot would have attributes such as characters, actions, locations, and weather. For a marketing strategy, you might use attributes of markets open to you, uses of the product, and skills you have available.

Next, draw up a table using these attributes as column headings. Write down as many variations of the attribute as possible within these columns.

This might be an exercise that benefits from brainstorming. The table should now show all possible variations of each attribute. Now select one entry from each column. Either do this randomly or select interesting combinations. By mixing one item from each column, you will create a new mixture of components. This is a new product, service, or strategy. Finally, evaluate and improve that mixture to see if you can imagine a profitable market for it.

For example, imagine that you want to create a new lamp. The starting point for this might be to carry out a morphological analysis. Properties of a lamp might be power supply, bulb type, light intensity, size, style, finish, material, shade, and so on. The core idea is to *forget* everything we have previously learned and retained when thinking about a possible solution.

We will first have to *break up* an existing product, service, or system *into its elements*. We have to ask: "Is this element of an object or a structure necessary for its function?" The elements have certain characteristics/attributes (form, material etc.). Are there any other possibilities and ways to create this element? Ask: what happens when we put elements with changed attributes together?

For this purpose, the elements and possible attributes are listed in Table 3.1. We might arrive at surprising *compositions* from the table; not all will work but some of them will lead to completely new products, services, and structures. As shown below, you can set these out as column headings on a table, and then brainstorm variations. Table 3.1 is sometimes known as a *morphological box* or *Zwicky box*.*

Interesting combinations might be as follows:

- Solar powered/battery, medium intensity, daylight bulb—possibly used in clothes shops to allow customers to see the true color of clothes.
- Large hand-cranked arc lights—used in developing countries, or far from a main power supply.
- Ceramic oil lamp in Roman style—used in themed restaurants, resurrecting the olive oil lamps of 2000 years ago.
- Normal table lamp designed to be painted, wallpapered, or covered in fabric so that it matches the style of a room perfectly.

---

* Named after the scientist Fritz Zwicky, who developed the technique in the 1960s.

TABLE 3.1

Zwicky Box

Power Supply	Bulb Type	Light Intensity	Size	Style	Finish	Material
Battery	Halogen	Low	Very large	Modern	Black	Metal
Mains	Bulb	Medium	Large	Antique	White	Ceramic
Solar	Daylight	High	Medium	Roman	Metallic	Concrete
Generator	Colored	Variable	Small	Art nouveau	Terracotta	Bone
Crank			Handheld	Industrial	Enamel	Glass
Gas				Ethnic	Natural	Wood
Oil/petrol					Fabric	Stone
Flame						Plastic

Some of these might be practical, novel ideas for the lighting manufacturer. Some might not be. This is where the manufacturer's experience and market knowledge are important. Morphological analysis, matrix analysis, and attribute listing are useful techniques for making new combinations of products, services, and strategies. You use the tools by identifying the attributes of the product, service, or strategy you are examining.

Attributes might be components, assemblies, dimensions, color, weight, style, speed of service, skills available, and so on. Use these attributes as column headings. Underneath the column headings, list as many variations of that attribute as you can. You can now use the table or *morphological box* by randomly selecting one item from each column, or by selecting interesting combinations of items. This will produce ideas that you can examine for practicality.

The standard steps of the process are as follows:

1. Identify the product/process/service or the components of the product/process/service you are dissatisfied with or wish to improve.
2. Make a list of elements of a product or a service or a list of elements of an organizational strategy of the company.
3. List all of the elements (e.g., material, color, weight, use of the product, and design) that can be described with certain attributes.
4. Choose some of these attributes that seem particularly interesting or important.
5. Identify alternative ways of achieving each attribute, either by conventional inquiry, or via an idea-generating technique, for example, brainstorming.

6. Combine one or more of these alternative ways of achieving the required attributes and try to come up with a new approach to the product or process being worked on. For example, the combination of a solar power system with a modern style bulb, wood for pole material, medium bulb size, and average light intensity with the pole fixed to the ground.

7. Discuss the feasibility of implementing these alternatives.

The following questions might also be useful when applied to a situation or problem: Adapt? Modify? Replace with? Magnify/maximize? Rearrange? Reverse? Combine? Minimize/remove? Add/subtract something? Change color? Vary materials/reorganize parts? Change shape? Adapt style?

SUMMARY

Attribute listing focuses on the attributes of an object, seeing how each attribute could be improved. *Morphological analysis* uses the same basic technique, but is used to create a new product by mixing components in a new way. Attribute listing can also be used in business situations such as marketing and product positioning. The position of a product in the market may be based on the following factors:

- Product features—organically grown vegetables
- Product benefits—a particular car model has additional space in back, which can be changed to hold additional seats or extra storage space and is ideal for families
- Associating the product with a use or application—the *honeymoon destination*
- User categories—association of the product with a user or class of users, for example, mobile phones for the young executive
- With respect to competition—compatible with *X* product

However, the position of a product may need to change or be adjusted for a variety of reasons, that is, new competition, new technology, new customer preference, etc. Attribute listing could be used in this situation to find a new or adjust the existing position of a product in the market.

Advantages

Every new product, program, or service results from an idea that then follows an innovation cycle of testing, implementing, and marketing. It is generally accepted that the key to competitive advantage is generating and successfully exploiting new ideas that come from creative thinking of both individuals and groups.

Disadvantages

The environment in some organizations may prove to be hostile to creative thinking. Anyone managing this creative thinking process is likely to encounter obstacles such as (a) general resistance to change, (b) free expression being stifled by a pervasive culture of blame, (c) a failure being regarded as a cause of penalties and not an opportunity to learn, (d) rigid formalities and rules, (e) inadequate and nonexistent incentives leading to slow decision making, (f) reluctance to think and move outside of strict job descriptions, and (g) a view that the best ideas come from the top.

EXAMPLES

1. Robert Harris, in his article "Creative thinking techniques," gave the following example of how attribute listing can be used to solve a problem:

   For example, let's say you work for a ball bearing manufacturer and you discover that a flaw in one of the machines has caused the production of 800 million slightly out-of-round ball bearings. You could ask, "What can I do with 800 million slightly out-of-round ball bearings?," and, of course, a few things come to mind, like sling shot ammo and kid's marbles. But you could also break the ball bearings down into attributes, such as roundish, heavy, metal, smooth, shiny, hard, and are able to be magnetized. Then you could ask, "What can I do with 800 million heavy things?," or "What can I do with 800 million shiny things?"

   You can focus on each identified attribute and ask questions about it, like this: What can heavy things be used for? Paperweights, ship ballast, podium anchors, tree stands, scale weights, and so on. What can be done with metal things? Conduct electricity, magnetize them, melt them, and make tools with them.

TABLE 3.2

Attribute Listing

Features	Attributes	Alternatives
Material used	Wood	Plastic
Lead material	Gray	Luminous
Color	Brown	Multicolor with company logo
Shape	Cylindrical	Oval
Special features	Eraser on top	Hook to clip onto clipboards

2. Another example of attribute listing is provided by Innosupport,* using the components or attributes of a pencil, which are

- Shaft material
- Lead type
- Hardness and width of lead
- Quality
- Color
- Weight
- Price, etc.

   With attribute listing, you list all of the alternatives to these components and mix the alternatives generated to find a new innovative product or service (Table 3.2).

3. Another similar example is a flashlight, which may be described as a long, round tube made of plastic using batteries to light a bulb, which shines through a clear plastic shield when the user pushes a switch. Examining each of the attributes could lead to new ideas, as suggested in the pencil example above. Why is it round? Why plastic or metal? Could it be turned on in a different way, or be powered by a different source? This kind of questioning could lead to new products that would address entirely new markets.

* See http://www.innosupport.net/index.php?id=2024 for more details.

CASE STUDY

Clinical Imagination—Dynamic Case Studies Using an Attribute Listing Matrix

The Attribute Listing Matrix Case Study (ALMCS) is an active instructional strategy for use in the classroom or clinical laboratory designed to engage the learner at the analysis, synthesis, and evaluation levels of Bloom's cognitive domain. Random numbers are used to generate multiple versions of case studies within a matrix that contains categories of real-world variables. Nursing students, either individually or in small groups, can then use the nursing process to analyze the patient case and design an individualized care plan. The ALMCS can be readily adapted to any level of nursing education and to any clinical specialty. It can be used in the classroom to show students how they will apply theoretical knowledge to real clinical situations, or it can be used for summative assessment by generating a random case for students to respond to in an examination.

For details, see http://www.ncbi.nlm.nih.gov/pubmed/21710962.

SOFTWARE

- *iThinkerBoard for the iPad and ThinkPack for the card deck.* Looking for a unique invention, an untapped market for an existing product, or a new solution? Stretch and flex your mental muscles with *iThinkerBoard* and a card-deck version, ThinkPak—creative-thinking tools that provide a unique way to organize and visualize your ideas. You can use it as a life planner, project management tool, or task organizer. This is made for the iPhone, and the possible applications are to expand your ideation by enhancing a method of stimulation of your creative power. ThinkPak is 56 individual cards used to create new and innovative ideas. Not only can the cards be used individually but also with groups, co-workers, teammates, family, children, etc. If you have not read the books *Thinkertoys* or *Cracking Creativity*, these tools are merely a way to create new ideas.
- *The Creative Thinker by Idon Resources.* This software brings you non-linear, yet structured, thinking with the ability to further develop your thoughts in the form of user-friendly and fun-to-use graphical hexagon

modeling for attribute listing. You can even organize your knowledge, supplement ideas with notes, and directly and meaningfully link ideas to the Internet, documents, slide presentations, spreadsheets, and more. The Creative Thinker literally allows you to visualize your thoughts, rapidly access knowledge, and combine the two creatively in real time for insights. See http://www.idonresources.com/ct/creativethinker.html.

- *Creative thinking program applied to attribute listing.* This program can also be customized for attribute listing. The objective is to use the attribute listing tool in combination with other innovation methods to drive innovation, improvisation, and creative thinking into a team and organization. A combination of skill development training, real-time process facilitation, team and individual coaching, and interviews are usually used by the innovation coach. Programs include a repeatable creative thinking for attributes process that can be used by different teams, along with relative thinking best practices, tools, and techniques for individuals and teams (http://www.creativeemergence.com/creativethinking.html).

SUGGESTED ADDITIONAL READING

Erl, T. *SOA: Principles of Service Design.* Upper Saddle River, NJ: Prentice Hall, 2008.

Gitzel, R., Korthaus, A., and Schader, M. Using established web engineering knowledge in model-driven approaches. *Science of Computer Programming,* vol. 66, no. 2, pp. 105–124, 2007.

Higgins, J. *101 Creative Problem Solving Techniques: The Handbook of New Ideas for Business.* Revised edition. Winter Park, FL: New Management Pub Co., 2005.

Mauzy, J. and Harriman, R. *Creativity Inc.: Building an Inventive Organization.* Boston: Harvard Business Review Press, 2003.

Nolan, V. *The Innovator's Handbook.* London: Sphere Books, 1989.

Prince, G.M. *The Practice of Creativity: A Manual for Dynamic Group Problem-Solving.* Brattleboro, VT: Echo Point Books & Media, LLC, 2012.

Roth, W. and Voehl, F. *Problem Solving for Results.* Delray, FL: St Lucie Press, 1998.

Smolensky, E.D. and Kleiner, B.H. How to train people to think more creatively. *Management Development Review,* vol. 8, no. 6, pp. 28–33, 1995.

Toll, M. *What is PED?*, https://www.linkedin.com/in/marvintoll, and Pattern Enabled Development® (PED) (http://pedCentral.com), 2012.

WEBSITES

http://www.mycoted.com/Attribute_Listing
http://members.tripod.com/~eng50411/psolving.htm

http://www.streetdirectory.com/travel_guide/2570/family/the_elements_of_creativity
_attributes_listing_method.html
http://for-creativity.blogspot.com/2007/09/elements-of-creativity-attributes.html
http://www.cmctraining.org/articles_view.asp?article_id=55&sid=0
http://www.mindtools.com/pages/Reviews/CreatingNewProductsandServices.htm
http://www.eirma.org/members/learninggroups/lg03-creat/Learngroup_creat_follow-up
/Ramon-Vullings-creativity_tool_catalogue.pdf
http://www.4pm.pl/page/37/index.html
http://road.uww.edu/road/zhaoy/old%20course/Chapters%209&11.ppt

4

Brainwriting 6–3–5

Achmad Rundi

CONTENTS

Nothing is more dangerous than an idea, when it's the only one we have.

Emile Chartier

DEFINITION

Brainwriting 6–3–5 is an organized brainstorming with writing technique to come up with ideas in the aid of innovation process stimulating creativity. Professor Bernd Rohrbach developed this type of brainwriting and had it originally published in volume 12 of the German sales magazine *Absatzwirtschaft* in 1969 (6–3–5 BrainWriting, n.d.).

It is derived from the brainstorming process. It is used to generate ideas within a group setup. The four principles of brainstorming are as follows: focus on quantity, avoid criticism, any and unusual idea are welcome, and combine and improve ideas (Method 6–3–5 and 6–3–9, n.d.).

USER

This tool can be used by individuals or groups, but its best use is with a group of four to eight people. Cross-functional teams usually yield the best results from this activity.

OFTEN USED IN THE FOLLOWING PHASES OF THE INNOVATIVE PROCESS

The following are the seven phases of the innovation cycle. An X after the phase name indicates that the tool/methodology is used during that specific phase:

- Creation phase X
- Value proposition phase
- Resourcing phase
- Documentation phase
- Production phase X
- Sales/delivery phase
- Performance analysis phase

TOOL ACTIVITY BY PHASE

- Creation phase—During this phase, brainwriting facilitates quicker and more idea generation. It addresses the issues of distraction that can take place in a brainstorming done orally.
- Production phase—During this phase, brainwriting is frequently used to focus on solving production problems.

HOW TO USE THE TOOL

Brainwriting facilitates quicker and more idea generation. It addresses the issues of distraction that can take place in a brainstorming done orally. It minimizes conflicts/distractions such as group bias where a single person or a few can dominate the discussion, putting introvert people at a disadvantage in voicing their idea (6–3–5 BrainWriting, n.d.). 6–3–5 Brainwriting can also be carried out in a large room full of participants. The moderator's role is only to divide the groups and time the rounds of this method. The person administering the exchange with a large number of participants will form many sets of small groups and be able to finish the 6–3–5 brainwriting method in a short time. Using this tool is also suitable for a culture that does not allow exchange in a more open way. The writing part is to facilitate idea exchange in alignment with such culture but not discourage the possibility of idea collection.

The process of brainwriting 6–3–5 will take a group of six persons who will write three ideas based on a problem statement within 5 minutes. Afterward, the person will pass it to the other participants so the next participant can come up with the next three ideas based on the problem and stimulated by the ideas written down by the previous participants. At the end of the 30-minute process, there will be 108 ideas.

It is good that this process of ideation comes after we obtain a clear insight or have done the process of rooting out the problem. With this ideation step, we can generate ideas that can address the problems at hand. Then, however, it will require another tool to narrow down these ideas, such as a diverge and converge method. This brainwriting 6–3–5 tool will help generate the multiple ideas required in the collaboration.

An affinity diagram tool will help organize the 108 generated ideas (Wilson, 2013). It will group the ideas on the basis of their similarity, and this will help find themes to the idea we have collected. Please refer to the affinity diagram tool in Chapter 2 for more information.

Prerequisites: Have participants understand the background of the problem that will be the focal point of the brainwriting. Additionally, there is a need for a clear problem statement (Stacey, 2014).

PROCESS (STACEY, 2014)

Step 1: Establish a group of six people. They are sitting nearby so they can easily pass their paper after each round of 5 minutes of idea writing. They need to write neatly. They also need to state their idea clearly and concisely, and relevant to the problem statement. The group also states that the pass rotation will be consistent to the left or to the right. A circle formation will help.

Step 2: On the first round, each user will write three ideas based on the problem statement. A time keeper will keep the time of the first round to 5 minutes. At the end of the first round, the participants will pass the paper to the next participant in the order that was agreed on earlier.

Step 3: The second round will have the participants read the written ideas as inspiration or stimulation for their next three ideas. There is no discussion between participants while the 30 minute brainwriting is in session.

Step 4: Continue the rotation for the third, fourth, and up to the sixth round. Make sure each participant adds three ideas to every round. Each participant's sheet will have 18 ideas. For six sheets, it will be 6 × 18 = 108 ideas.

TABLE 4.1

Table Sample

Problem: How can we improve the knowledge of English in the service department?			
A1: Evening classes	A2: Motivation for holiday abroad	A3: "English speaking day" each week in the department	A: Nikos
B1: Evening classes paid by the company	B2: Motivation for summer university in England	B3: Project meetings in English	B: Vlad
C1: Evening classes with 2 hours off to prepare for the classes	C2: Contribution by the company for summer university in GB	C3: Conversation circle once a month after work	C: Annelie
D1: Two hours per week with a teacher in the company	D2: Higher remuneration for a certified qualification	D3: English version of professional literature	D: Linda
E1: e-Learning with a tutor and classes in the company	E2: English "library" at the company	E3: Conversation circle with foreign colleagues from the company	E: Laila
F1: Intensive course twice a year on the weekend	F2: Going to the Irish pub as often as possible	F3: Intercultural evenings with foreign colleagues from the company	F: Caron

Source: http://media.nuas.ac.jp/~robin/zakka/z-lesson/369.ht1.GIF.

EXAMPLE

See Table 4.1 for an example of brainwriting.

SOFTWARE

Nowadays, people have computers or tablets connected to a WiFi network within the room such as a conference or meeting room. Furthermore, those participants are in close range with each other. A collaborated e-document allows them to update the same file easily either simultaneously or in turn. With the similar process that the physical sheet can do, the shared electronic spreadsheet can collect the information the same way:

- Google Spreadsheet
- Office 365 Excel

REFERENCES

6–3–5 Method (BrainWriting). Retrieved from Youtube.com: http://www.youtube.com/watch?v=p-I6a6AqDBM, uploaded by linkmv97, April 28, 2009.

Method 6–3–5 and 6–3–9. Retrieved from Nagoya University of Art and Science: http://media.nuas.ac.jp/~robin/zakka/z-lesson/369.htm, n.d.

Stacey. 08 Ideas in 30 minutes—The 6–3–5 method of brainwriting. Retrieved from Blogsession: http://blogsession.co.uk/2014/03/635-method-brainwriting/, March 25, 2014.

Wilson, C. Using Brainwriting for Rapid Idea Generation. Retrieved from SmashingMagazine.com: http://www.smashingmagazine.com/2013/12/16/using-brainwriting-for-rapid-idea-generation/, December 16, 2013.

5

Cause-and-Effect Diagram

H. James Harrington

CONTENTS

Every negative effect is caused by something. The trick is to find the right cause so it can be corrected.

H. James Harrington

DEFINITION

A cause-and-effect diagram is a visual representation of the possible causes of a specific problem or condition. The effect is listed on the right-hand side and the causes take the shape of fish bones. This is the reason it is sometimes called a *fishbone diagram* or an *Ishikawa diagram*.

USER

This tool can be used by individuals, but its best use with a group of four to eight people. Cross-functional teams usually yield the best results from this activity.

OFTEN USED IN THE FOLLOWING PHASES OF THE INNOVATIVE PROCESS

The following are the seven phases of the innovative cycle. An X after the phase name indicates that the tool/methodology is used during that specific phase.

- Creation phase X
- Value proposition phase
- Resourcing phase
- Documentation phase
- Production phase X
- Sales/delivery phase
- Performance analysis phase

TOOL ACTIVITY BY PHASE

Cause-and-effect diagrams can be used whenever a problem is identified, and as a result could be used in all the phases. They are primarily used during the production phase and creation phase.

HOW TO USE THE TOOL

A cause-and-effect analysis is a structured analysis used to separate and define causes. The effects are the symptoms, which let us know that we have a problem.

The cause-and-effect diagram is very applicable to repetitive processes. For this reason, most of the problems can be analyzed using cause-and-effect analysis. Processes under the control of one group or organization and where responsibilities are clearly defined are also good candidates for cause-and-effect analysis.

Not only are there several types of cause-and-effect diagrams, but there are also several methods used to develop them. The *random method* is so called because members of the problem-solving group may suggest causes, which apply to any of the major subdivisions of the diagram. Typically, just as in brainstorming, the session has a leader and scribes are appointed to record the contributions of the members of the group. With the *systematic method* of developing a cause-and-effect diagram, the leader chooses one of the major subdivisions on which to focus the group's attention. The brainstorming process addresses the subdivision indicated by the leader. When that particular subdivision has been completed, the leader indicates the next one, and so on, until the cause-and-effect diagram has been systematically completed. The completed diagrams, whether generated by the random method or the systematic method, look alike.

Cause-and-effect diagrams have been variously called *fishbone* diagrams, as suggested by the shape of the completed diagram; *Ishikawa diagrams*, named after Professor Kaoru Ishikawa; and *cause diagrams*.

The process analysis diagram is another type of cause-and-effect diagram. It looks much like a flowchart.

The solution analysis diagram, in some ways, is much like the cause-and-effect diagram except that it can be considered a backward fishbone.

Constructing a Cause-and-Effect Diagram

Constructing a cause-and-effect diagram is a three-step process:

- Step 1. Name the problem or *effect*. This effect is placed in a box on the right and a long process arrow is drawn pointing to the box.
- Step 2. Decide on the major categories or subdivisions of causes. These major categories are placed parallel to and some distance from the main process arrow. Arrows slanting toward the main arrow then connect the boxes.
- Step 3. Brainstorm for causes. These causes are written on the chart, clustered around the major category or subdivision, which they

influence. Arrows pointing to the main process arrow connect them. The causes should be divided and subdivided to show, as accurately as possible, how they interact (Figure 5.1).

To have a little fun, use the cause-and-effect diagram shown in Figure 5.2 to perform the same analysis (Harrington and Lomax, 2000).

Remember that either the random method or the systematic method may be used to generate the fishbone diagram.

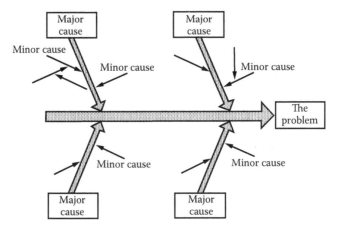

FIGURE 5.1
Cause-and-effect diagram.

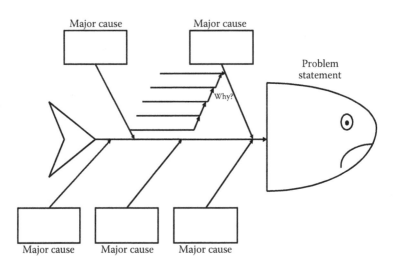

FIGURE 5.2
Cause-and-effect diagram.

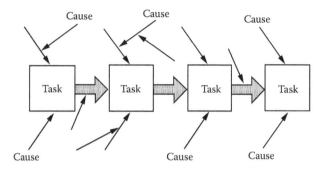

FIGURE 5.3
Process analysis diagram.

Constructing a Process Analysis Diagram

To construct the process analysis diagram (see Figure 5.3), it is first necessary to list the series of tasks you wish to analyze. List them in the order in which they are done. Next, place each of these in a box, and line the boxes up in proper order, connecting each of the boxes with arrows to indicate the progress of the process. The third step is to brainstorm, using either the random or systematic method, all of the causes that contribute to each step. These are listed on the chart and connected by arrows to the box containing the step to which they refer.

Constructing a Solution Analysis Diagram

With the cause-and-effect diagram and process analysis diagram, we start with the effect (problem) and analyze for causes. In the solution analysis diagram (see Figure 5.4), we do the reverse. Start with a single cause (a proposed solution) and analyze for all of the possible effects. Whether you have one or several possible solutions, you may wish to analyze them all. To best determine all the effects of a possible solution, use a solution analysis type of cause-and-effect diagram. It will help make better solution choices and answer questions about your activities (Harrington and Lomax, 2000).

The first step is to name the proposed solution, and enclose it in a box. The box is placed on the left side with a process arrow leading away from it. This arrow shows the direction of influence. Next, decide on the major categories or subdivisions of *effects*. (These are the areas the proposed solution is likely to influence.) The third step is to brainstorm, using either the systematic or random method, the likely effects (good and bad) of the proposed solution. These effects are clustered around the major categories

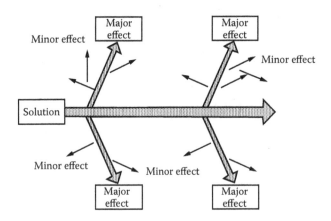

FIGURE 5.4
Solution analysis diagram.

with smaller arrows leading away from the arrows indicating major categories. After this step has been completed, it is good to list the positive outcomes in one column and the negative in another. Compare the positive and negative results of the proposed solution. Use this information in deciding which solution to employ.

Use of Cause-and-Effect Diagrams

There are several uses for cause-and-effect diagrams. They can be used to

- Assist both individual and group ideation
- Serve as a recording device for ideas generated
- Reveal undetected relationships
- Investigate the origin of a problem
- Investigate the expected results of a course of action
- Call attention to important relationships

Another aspect of diagramming causes and their effects is that one can tell at a glance whether the problem has been thoroughly investigated. A cause-and-effect diagram, which contains much detail, indicates how deeply a group has gone into the process of investigation if that detail is legitimate. On the other hand, a bare cause-and-effect diagram might indicate that the problem was not significant or that the solvers of the problem were not exhaustive in their search. Likewise, if the solution analysis diagram is complete, it will show the group's concern for the impact of a proposed solution.

Guidelines for Constructing Cause-and-Effect Diagrams

When constructing a cause-and-effect diagram, you should give attention to a few essentials that will provide a more accurate and usable result (Harrington and Lomax, 2000).

1. Participation by everyone concerned is necessary to ensure that all causes are considered. All members must feel free to express their ideas. The more ideas mentioned, the more accurate the diagram will be. One person's idea may trigger someone else's.
2. Do not criticize any ideas. To encourage a free exchange, write down all ideas just as they are mentioned in the appropriate place on the diagram.
3. Visibility is a major factor of participation. Everyone in the group must be able to see the diagram. Use large charts, use large printing, and conduct diagram sessions in a well-lighted area.
4. Connect related causes as they are mentioned so the relationships can be seen as the diagram develops.
5. Do not overload any one diagram. As a group of causes begins to dominate the diagram, that group should be isolated and a separate diagram made for those causes.
6. Construct a separate diagram for each separate problem. If your problem is not specific enough, some major categories of the diagram will become overloaded. This indicates the need for additional diagrams.
7. Circle the most likely causes. This is usually done after all possible ideas have been posted on the cause-and-effect diagram. Only then is each idea critically evaluated. The most likely ones should be circled for special attention.
8. Create a solution-oriented atmosphere in each session. Focus on solving problems rather than on how problems started. The past cannot be changed—only the future can be affected by eliminating causes of undesired effects.
9. Understand where each cause is to be placed on the diagram.

EXAMPLES

Figure 5.5 shows two typical examples cause-and-effect diagrams.

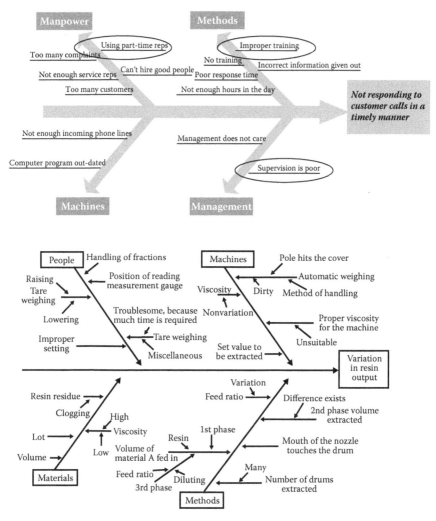

FIGURE 5.5
Typical examples of cause-and-effect diagrams.

SOFTWARE

Some commercial software available includes but is not limited to

- Smartdraw: www.smartdraw.com/
- Edraw: www.facebook.com/edrawsoft
- QI macros: http://www.qimacros.com

REFERENCE

Harrington, H.J. and Lomax, K. *Performance Improvement Methods*. New York: McGraw-Hill, 2000.

SUGGESTED ADDITIONAL READING

Asaka, T. and Ozeki, K., eds. *Handbook of Quality Tools: The Japanese Approach*. Portland, OR: Productivity Press, 1998.
Brassard, M. *The Memory Jogger Plus*. Milwaukee, WI: ASQ Quality Press, 1989.

6

Creative Problem-Solving Model

Dimis Michaelides

CONTENTS

DEFINITION

Creative problem solving (CPS) is a methodology developed in the 1950s by Osborn and Parnes. The method calls for solving problems in sequential stages with the systematic alternation of divergent and convergent thinking. It can be enhanced by the use of various creative tools and techniques during different stages of the process.

USER

This method can be used by individuals, teams, and organizations to confront a wide range of complex challenges—business, social, and personal. CPS can be applied to any problem (challenge) that is open to creative investigation and resolution.

OFTEN USED IN THE FOLLOWING PHASES OF THE INNOVATIVE PROCESS

The following are the seven phases of the innovative cycle. An X after the phase name indicates that the tool/methodology is used during that specific phase.

- Creation phase X
- Value proposition phase X
- Resourcing phase X
- Documentation phase X
- Production phase X
- Sales/delivery phase X
- Performance analysis phase

TOOL ACTIVITY BY PHASE

The CPS model is one of those universal tools that can be used in all phases. It is effectively used whenever a problem is identified in order to find a creative way of solving that problem. Unfortunately, in all of the first five phases of the innovative cycle, a number of problems can arise that would benefit from the use of this tool.

HOW TO USE THE TOOL

Background

CPS was developed by Alex Osborn of BBDO, an advertising firm, and Sidney Parnes of the University of Buffalo. Alex Osborn is also credited as the originator of *brainstorming* because he codified its practice and set it within the context of CPS. Sidney Parnes was a professor of creative studies who founded the first recognized department of creative studies at the university level.

Method

The basic premise of CPS is that many complex problems have more than one solution. Such problems are best approached in sequential stages with the systematic practice at each stage of two fundamentally different types of thinking: divergent and convergent.

Sequential Stages

The sequential stages should be at least three:

1. *Preparation:* to understand and define the problem.
2. *Solution:* to resolve the problem.
3. *Implementation:* to make the solution happen.

The above three *universal* stages can be further broken down. Initially, Osborn and Parnes began with a four-stage model, which they later extended to a six-stage model.

Divergence and Convergence

What makes the CPS method unique is the systematic alternating practice of divergent and convergent thinking. This came from the realization that creative (divergent) thinking and critical (convergent) thinking, which are both extremely useful for CPS and for humanity in general, should therefore be practiced separately—first divergence, then convergence—because they call for significantly different working norms.

Divergence is the search for alternatives and is subject to four *norms* or *rules*:

- *Suspend judgment.* There should be no critique or judgment of any kind during this phase. While simple, this norm is very different from our usual ways of functioning at work.
- *Quantity.* It is important to generate many facts/problem definitions/solutions, etc., to create a wide array of alternatives for consideration. Practice also suggests that originality often turns up later, after more conventional ideas/solutions have been considered.
- *Beyond reason.* A conscious effort is required to stretch the mind to the limits of the possible, into the impossible, the world of fantasy, magic, unreason, and to record such way-out expressions. This is supported by the fact that many great discoveries have not had logical origins. Alex Osborn argued that it is easier to tame a wild idea than it is to make a boring one more interesting.
- *Build on other ideas.* Once an idea is on the table, it can be enriched and transformed into more and different ideas.

Divergence will inevitably end up in long lists and no clear conclusion, which is why the process is incomplete without *convergence.*

Convergence is about making choices from the options provided in the convergence phase and is subject to the following *norms* or *rules*:

- *Classify.* Ordering materials in clusters of similar items to organize the output of divergence.
- *Evaluate.* This is where judgment is reintroduced—weighing the pros and cons of each item and prioritizing the most attractive options.
- *Select.* Choices must be made among available options before moving on.

Divergence and convergence are both active processes requiring process management and personal skills. Skillful divergence will allow for both individual and team sessions. Skillful convergence will strive to keep a positive spirit and not stifle originality.

Method—Bringing Together Sequential Stages, Divergent and Convergent Phases

We here describe the I.D.E.A.S! model, an application of the Osborn–Parnes method developed by the author of this chapter which has *five stages*, within each of which *two phases* are actively practiced—divergence and convergence. Various techniques are useful at the different stages of the process, and the ones mentioned below are only a few among many available.

- I—Investigate.
  During this stage, facts are collected, and assumptions and feelings are exchanged to create a basis for the challenge.

  Useful techniques for the divergent phase are picture association, word association, divergent mind mapping, and metaphors, analogies and, in fact, any technique that encourages an indirect look at the problem through an *exit* from the topic followed by a *reentry*.

  In the convergent phase, the most important facts and feelings are agreed on and summarized.

  By the end of this stage, a manageable number of the most important facts (say 10–20) and a *mini-vision* (say two sentences giving an image of the future to be achieved after the problem is resolved) have been defined.
- D—Define.
  This is a crucial stage since, following the Charles Kettering's dictum, "a problem well stated is a problem half solved." During this stage, the problem is seen from different perspectives and then framed, clearly and succinctly.

  The divergent phase should be devoted to reformulating or paraphrasing the initial problem through open questioning—How might we...? In what ways might I ...? This enables examination of the problem from many different viewpoints. Questions can be derived from facts and aspirations expressed in the previous—Investigate—stage plus other useful sources, known or novel. Stretching the imagination to *impossible* questions will often yield insights that are out of the ordinary. Brainstorming as many questions as possible is useful here as are all the associative techniques quoted in the Investigate stage. Quite often the definitions will be linked, each describing different levels of generality and specificity.

In the convergent phase, it is useful to cluster the different questions/problem definitions in logical categories (convergent mind mapping may be useful here), and then serious choices must be made. A useful technique is the three I's selection technique: Importance—how relevant is the problem thus defined? If resolved, how likely is it to bring considerable value? Imagination—how open to creative discovery is it (so it is not simply pointing to a solution everyone already knows)? Influence—how likely are we to have the ability to implement solutions, that is, are actions likely to be within our authority and control? It is possible at this stage to agree on two or three (or even more) *subproblems* to work on, because a complex problem may be best solved when broken into smaller, more manageable components. In this case, the following stages are best applied to each of the subproblems separately or in parallel, with intelligent project management ensuring coherence between the parts.

By the end of this stage, there is a clear and simple definition of the problem, or a small number of subproblems.

- E—Envision.

This is the pure solution-finding stage of the process, in which we imagine different ways of resolving the problem.

Many techniques are available for the divergent phase. Brainstorming is good as an opener. It can be complemented with associative techniques (such as analogies), combinatory techniques (such as What if?), or dream-sourced techniques (such as image streaming) to discover original solutions. Provocation and magical idea techniques, which open the door to wild fantasy before ensuring a safe landing, are also likely to greatly stretch the imagination.

Convergence during this stage calls for selecting the best ideas. This requires both silent contemplation and thoughtful, constructive discussion. It is good to cluster ideas into Wow! (very inspiring, highly original, high value-adding), Now (easy to implement, slightly value adding), How? (impossible ideas that will need a lot of tweaking to turnaround into feasible solutions). Thereafter, a long list and then a short list of the best solutions should be drawn up.

By the end of this stage, there is a sizeable short list (of, say, 10–15 solutions) that includes Wow! How? and Now solutions.

- A—Appraise.

Up to this stage of the process, solutions generated have not been analyzed or understood in any depth. New ideas are still at an undeveloped, infant stage. This is the stage at which ideas are examined in some depth, developed, and strengthened to stand up as possible, valuable, and feasible solutions, and then the best one(s) is (are) selected for implementation.

Divergence at this stage consists of examining each short-listed idea in turn. The pluses, potentials, concerns, overcome (PPCO) technique is very useful here. *Pluses* indicate the advantages of the solution. *Potentials* indicate what else might be done with this solution or to what additional things might it lead to. *Concerns* list the disadvantages or difficulties that might be encountered. *Overcome* lists the ways of addressing the concerns. Practicing PPCO requires skill, imagination, and thoroughness, as well as suspending judgment. Each of the four headings should generate a rich list, after which it is possible to write up a good description of the solution.

The convergence phase calls for a choice of the best solution(s). Good criteria are value (benefit–cost–risk assessment), feasibility, and originality. A final decision may require the investment of additional resources and expertise to make a good business case.

By the end of this stage, the best solution has been chosen for implementation.

- S!—Start!

During this stage, a project will be brought to fruition through the implementation of a solution to a problem.

During the divergent phase, many possible ways of making the project happen are examined.

During the convergent phase, choices are made between alternative courses of action, and ways of monitoring implementation are defined. By the end of this stage, an action plan has been agreed on, with clear action responsibilities, names, and deadlines.

CPS—Comments and Points of Practice

CPS model. It is interesting to note that in this model, only one out of five stages is devoted to finding solutions. Two stages are preparatory (Identify,

Define) and two stages are about developing and implementing solutions (Appraise, Start!). The discussions that happen during the preparatory stage are extremely useful, and jumping to solutions prematurely may waste a lot of creative energy. Making ideas happen also requires thought and debate, and moving to action prematurely may limit the generation of novel alternatives and in-depth appraisal.

CPS should never be practiced mechanically. If at times the outcome of a stage is not satisfactory, then taking one or two steps back to an earlier stage may be very useful. New data may arise, new discoveries may be made on the way, and so such backtracking may be of value.

Active practice. Movement from one stage to another is deliberate. Without separation of stages, the beginning of problem solving is often a mix of facts, problems/challenges, and solutions. Movement from active divergence to active convergence is also deliberate. Without the separation of phases, there is often premature judgment and debate—thesis and antithesis—without the conscious introduction of novelty. Often proposal and judgment are happening at the same time, a practice that does not encourage novelty.

Who should practice CPS? Above all, people who *own* the problem and who have a stake in its resolution. It has been suggested by some practitioners (IDEO video clips) that those participating in the divergent phases should not be those doing the convergence. Such an approach might at times be useful, though it has its merits and drawbacks.

Time management. Managing time between stages and phases is a matter of skillful judgment. There is no golden rule for how much time to dedicate to each stage or each phase. Overall, time and effort will be more, the larger and more complex the problem. The quest for great originality in the outcome may often necessitate more time for Envision.

Professional facilitation and practice. As the sequence of stages and phases is different from conventional ways of working, professional facilitation is initially a necessity. As people get used to working the CPS way, they can take it in turns to facilitate. Virtual collaboration and sharing platforms can greatly help the process as many parts of the process can be done asynchronously.

Team building value. The norms of divergent thinking encourage collaboration, teamwork, and *buy-in. Suspending judgment* allows for non-confrontational interaction. *Quantity* calls for energy to produce many alternatives. *Beyond reason* calls for expression and communication at a level that is much deeper than reason alone. *Building on other ideas* calls for acceptance and positive enrichment of all ideas proposed by team members. This forms a good basis for the discussion and debate that must follow in the convergent phase.

Individual or teamwork? There is no doubt that many novel ideas are generated by individuals thinking alone or acting in solitary mode. There is also no doubt that most innovation is the outcome of creative collaboration that improves, transforms, and implements individual or group ideas. Solo work or teamwork will each have its drawbacks too. The best outcomes of CPS will be those that allow the practice of both, and at both the divergent and convergent phases. This will enrich both the individuals' contribution and the value added by team collaboration.

Creative styles. It has become clear that, while all people are creative, they use their creative skills in different ways. A number of researchers—Basadur (CPS Profile) and Puccio (Foursight)—have linked different individual personal profiles to different stages of CPS, clarifying why creative collaboration may be more or less challenging and emphasizing the importance of a diversity of profiles for innovation.

Creative leadership and organizational context. It is clear that the effectiveness of CPS in an organization will also be dependent on structures and cultures that encourage (or hinder) innovation (Michaelides, 2007).

Extensions, developments, and refinements. CPS has been redesigned and enriched by many different researchers and practitioners including the author (I.D.E.A.S! model), Min Basadur (Simplex), Tim Hurson (Think x), and Gerard Puccio (Foursight). The mathematician Jacques Hadamard also independently proposed a similar problem-solving method using five stages (Jaoui, 1994). These versions all preserve sequential stages and divergence–convergence.

Design thinking. Recent trends in design thinking use many principles of CPS such as a collaborative approach to problem solving, deep immersion in the problem in the early investigative phases, and the use of brainstorming for solutions. An important difference from the classic CPS model is the insistence of creating rapid prototypes, testing these, and dropping them or redesigning them as one moves along.

EXAMPLES

See the examples embodied in the section of this chapter entitled "How to Use the Tool."

SOFTWARE

No specific software is recommended to support this tool.

REFERENCES

Jaoui, H. *La Créativé Mode d'Emploi: Applications Pratiques.* 2nd Ed. Paris: ESF, 1994.
Michaelides, D. *The Art of Innovation—Integrating Creativity in Organizations.* Nicosia: Performa Productions, 2007.

SUGGESTED ADDITIONAL READING

Basadur, M. *Simplex: A Flight to Creativity.* Buffalo, NY: CEF Press, 1994.
Basadur, M. *Impacts and Outcomes of Creativity in Organizational Settings.* Hamilton, Canada: McMaster University, 1993.
Basadur, M. *Managing the Creative Process in Organizations.* Hamilton, Canada: McMaster University, 1994.
Hurson, T. *Think Better: An Innovator's Guide to Productive Thinking: (Your Company's Future Depends on It... and so Does Yours).* New York: McGraw-Hill, 2008.
Kelly, T. *The Ten Faces of Innovation: IDEO's Strategies for Beating the Devil's Advocate & Driving Creativity Throughout your Organization.* New York: Doubleday, 2005
Osborn, A.F. *How to Think Up.* 1942. Reprinted in S.J. Parnes, ed. *Source Book for Creative Problem Solving: A Fifty Year Digest of Proven Innovation Processes.* Buffalo, NY: CEF Press, 1992.
Osborn, A.F. *Applied Imagination: Principles and Procedures of Creative Problem Solving.* New York: Charles Scribners, 1953; 3rd rev. ed., Buffalo, NY: CEF Press, 1993.
Parnes, S.J., ed. *Source Book for Creative Problem Solving: A Fifty Year Digest of Proven Innovation Processes.* Buffalo, NY: CEF Press, 1992.
Parnes, S.J., ed. *Optimize the Magic of Your Mind.* Buffalo, NY: Bearly Ltd., 1997.
Parnes, S.J., ed. *Visionizing: Innovating Your Opportunities.* 2nd Ed. Buffalo, NY: CEF Press, 2004.
Puccio, G. and Murdoch, M., ed. *Creative Leadership: Skills that Drive Change.* Thousand Oaks, CA: Sage Publications Inc., 2007.

7

Design for X

Douglas Nelson

CONTENTS

Design for X is the key to competitive, profitable products.

DEFINITION

Design for X (DFX) is both a philosophy and methodology that can help organizations change the way that they manage product development and become more competitive. DFX is defined as a knowledge-based approach for designing products to have as many desirable characteristics as possible. The desirable characteristics include quality, reliability, serviceability, safety, user friendliness, etc. This approach goes beyond the traditional quality aspects of function, features, and appearance of the item.

USER

This tool can be used by individuals, but is best used with a group of four to eight people. Cross-functional teams usually yield the best results from this activity.

OFTEN USED IN THE FOLLOWING PHASES OF THE INNOVATIVE PROCESS

The following are the seven phases of the innovative cycle. An X after the phase name indicates that the tool/methodology is used during that specific phase.

- Creation phase X
- Value proposition phase X
- Resourcing phase
- Documentation phase
- Production phase
- Sales/delivery phase
- Performance analysis phase

TOOL ACTIVITY BY PHASE

- Creative phase—During this phase, the product is being designed and the designer must consider all the areas that are affected by the design. Is the design easy to assemble (design for manufacturability)? Will the output be easy to maintain (design for maintainability)?
- Value proposition phase—How the DFX factors were considered during the creative cycle greatly influences the value related to the product design.

HOW TO USE THE TOOL

AT&T Bell Laboratories coined the term DFX to describe the process of designing a product to meet the above characteristics. In doing so, the life cycle cost of a product and the lowering of downstream manufacturing costs would be addressed (see Figure 7.1).

The earliest stages of product realization have the greatest effect on life cycle costs, yet they represent the smallest proportion of life cycle expenditures.

Gattenby and Foo, 1990

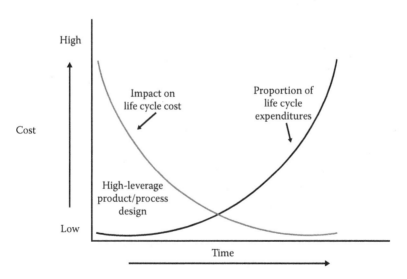

FIGURE 7.1
Design and life cycle costs.

Any number of factors might be relevant to the definition of quality during a systems design initiative. DFX involves being able to incorporate a variety of X factors into a design, working toward a solution set that optimizes their interaction against customer needs and requirements. Common X factor examples include assembly, reliability, and testing.

The DFX toolbox has continued to grow in number from its inception to include hundreds of tools. Some researchers in DFX technology have developed sophisticated models and algorithms. The usual practice is to apply one DFX tool at a time.

DFX tools may be incorporated as part of a larger concurrent engineering, product development, process reengineering, redesign, Six Sigma, or TRIZ (theory of inventive problem solving) project.

The benefits in the use of DFX that are directly related to competitiveness measures include improved quality, compressed cycle time, reduced life cycle costs, increased flexibility, improved productivity, more satisfied customers, safer workplace, and reduced environmental impacts. Time to market may be reduced, resulting in higher market share and increased profitability.

The use of DFX tools is likely to increase the number of design changes at early stages, but reduce the number of late design changes significantly. Because it is easier to change earlier rather than later, substantial savings can be achieved, allowing better products to be produced because products and processes may be realigned more easily and at lower cost. Some additional benefits of applying DFX involve operational efficiency in product development. These may include (a) better communication and closer cooperation, (b) concurrence and transparency, (c) earlier customer and supplier involvement, (d) improved project management, (e) enhanced team environment, (f) better structuring of product development processes, and (g) promotion of concurrent engineering.

It is advisable to use DFX tools as early as possible in the design or redesign process. The creation and value proposition phases can provide valuable inputs to the DFX project. Outputs from the DFX project may be useful in refinement and updating within the value proposition phase. As the innovative product and organization pass through the stages of financing, documentation, production, and sales/delivery, significant cost reduction, as well as quality and time improvements, may be realized through early investment in DFX tools. These benefits will be borne out within the performance analysis stage of the innovative process.

The tool is frequently used within concurrent engineering, and can be used in new designs or in subsequent redesigns. The specific tool selected will depend on the product iteration and priority of goals and objectives relative to product design and cost and time constraints.

Design Guidelines

DFX methods are usually presented as design guidelines. These guidelines provide design rules and strategies. The design rule to increase assembly efficiency requires a reduction in part counts and part types. The strategy is to verify that each part is needed or that common part types could be used.

A number of general DFX guidelines have been established to achieve higher quality, lower cost, improved application of automation, and better maintainability. Examples of these guidelines for design for manufacturing include the following:

- Reduce the number of parts to minimize the opportunity for a defective part or an assembly error, allowing potential decrease in the total cost of fabricating and assembling the product, and to improve the chance to automate the process.
- Mistake-proof the assembly design so that the assembly process is unambiguous.
- Design verifiability into the product and its components to provide a natural test or inspection of the item.
- Avoid tight tolerances beyond the natural capability of the manufacturing processes and design in the middle of a part's tolerance range.
- Design *robustness* into products to compensate for uncertainty in the product's manufacturing, testing, and use.
- Design for parts orientation and handling to minimize non-value-added manual effort, to avoid ambiguity in orienting and merging parts, and to facilitate automation.
- Design for ease of assembly by utilizing simple patterns of movement and minimizing fastening steps.
- Utilize common parts and materials to facilitate design activities, to minimize the amount of inventory in the system, and to standardize handling and assembly operations.
- Design modular products to facilitate assembly with building block components and subassemblies.
- Design considerations for serviceability into the product.

Design Analysis Tools

Each DFX tool involves some analytical procedure that measures the effectiveness of the selected tool. A typical design for an assembly procedure should provide for analysis of such elements as handling time, insertion time, total assembly time, number of parts, and assembly efficiency. Each tool should have some method of verifying its efficiency.

While each design tool is typically considered individually, exploration of synergies and trade-offs can be useful in optimal positioning of products along the product life cycle curve.

DFX Procedure (Huang, 1996)

Step 1. Product analysis—Information related to the product is collected. Bills of material (BOMs) are used to display product structure. Other types of product data can be easily correlated with the product BOM. It is useful to obtain the product hardware to examine and understand features.

Step 2. Process analysis—Process analysis is primarily concerned with the collection, processing, and reporting of process-specific and resource-specific data. Operation process and flow process charts are established.

Step 3. Measuring performance—The process and product interactions can be measured in terms of the relevant performance indicators within the specified DFX tool. Additional data collection and processing may be required.

Step 4. Benchmarking—The objective is to determine whether the subject process is good and what areas contribute to it. Benchmarking primarily involves setting up standards and comparing the performance measurements against the established standard. Individual and aggregate benchmarks may be established. Once the performance standards and measurements are available, the areas where performance measurements are below standards can be identified.

Step 5. Diagnosis—On the basis of performance measurement and benchmarking, a determination is made as to what is good and what is not. To solve problems, it is necessary to know their causes. A cause–effect diagram may be useful in determining major causes for a problem. Root cause analysis techniques may be used to further identify specific conditions related to the problem.

Step 6. *Advise on change*—Explore as many improvement areas as possible for each problem area. Brainstorming is a useful technique. Redesign of the product and processes are dependent on specific circumstances. Changes may take place to composition, configuration, and characteristics at different levels of detail. Product changes may be made across entire product ranges, working principles, concepts, structures, subassemblies, components, parts, features, or parameters. Process changes may be made across product lines, business processes, procedures, steps, tasks, activities, or parameters. Product and processes are closely interrelated. It is important to consider the interrelationships between the two when making a change to either products or processes. A *what if* analysis may be helpful.

Step 7. *Prioritize*—The analysis may reveal a large number of problem areas within the product and processes. There may be many causes and solution alternates for each problem. Prioritization is often required due to limited resources. Prioritization should be based on some form of measurement. A Pareto chart may be constructed and can be used to show the relative frequency of events such as products, processes, failures, defects, causes and effects, etc. Analysis of the Pareto chart should aid in prioritization of the problem areas.

Design for Product Life Cycle

Design for product life cycle is a cradle-to-grave approach (see Figure 7.2) DFX methodologies can be used in design, production, useful life, and, finally, end of product life.

There are currently hundreds of DFX tools that have been developed. A few of the more popular design tools are discussed herein.

Design for Safety

Design for safety requires the elimination of potential failure-prone elements that could occur in the operation and use of the product. The design should make the product safe for manufacture, sale, use by the consumer, and disposal or reuse. Failure modes and effects analysis (FMEA) and fault-tree analysis are often incorporated within design for safety. FMEA is a fundamental hazard identification and frequency analysis technique that analyzes all the fault modes of a given equipment item for its effects on both other components and the system. A fault-tree analysis is a hazard identification and frequency

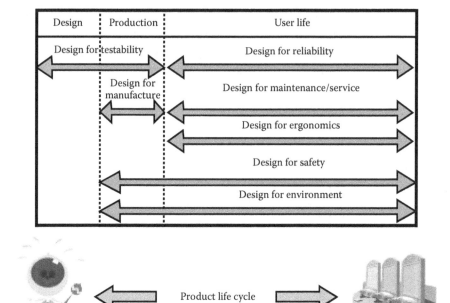

FIGURE 7.2
Design for product life cycle.

analysis technique that starts with the undesired event and determines all the ways in which it could occur. These are displayed graphically.

Design for Reliability

Design for reliability describes the entire set of tools that support product and process design (typically from early in the concept stage all the way through to product obsolescence) to ensure that customer expectations for reliability are fully met throughout the life of the product with low overall life cycle costs. A good tool to assess risk is the FMEA. The FMEA is used to identify potential failure modes for a product or process, assess the risk associated with those failure modes, prioritize issues for corrective action, and identify and carry out corrective actions to address the most serious concerns.

Design for Testability

Design for testability aims to make the product test procedures as easy and economical as possible during manufacturing, use, and servicing. Design for testability includes techniques that add certain testability features to a hardware product design. The idea behind design for testability is that features are added to make it easier to develop and apply manufacturing tests for the designed hardware. The purpose of manufacturing tests is to validate that the product hardware is free of manufacturing defects that may affect the product's correct functioning. Tests are applied at several steps in the hardware manufacturing flow and may also be used for hardware maintenance in the field. The tests generally are driven by test programs that execute within automatic test equipment processes. These tests may also be conducted within the assembled equipment during maintenance procedures. In addition to finding and indicating the presence of defects, tests may log diagnostic information about the nature of the subject test fails. The diagnostic information can be used to locate the source of the failure.

Design for Assembly/Manufacturing

Design for assembly means simplifying the product so that fewer parts are required, making the product easier to assemble and the manufacturing process easier to manage. Design for assembly is often the most effective DFX tool, providing increases in quality while reducing costs and time to market. Design for assembly is accomplished through the use of fewer parts, reduction of engineering documents, lowering of inventory levels, reduced inspection, minimization of setups, and materials handling. If a product contains fewer parts, it will take less time to assemble, reducing assembly costs. If the parts are provided with features that make them easier to grasp, move, orient, and insert, the assembly time and assembly costs will be reduced. The reduction of the number of parts in an assembly has the additional benefit of generally reducing the total cost of parts in the assembly. Major cost benefits of the application of design for assembly are achieved through this reduction of number of parts and, thus, cost of parts.

Design for Environment

Design for environment aims to create minimal levels of pollution over the product life cycle. Manufacture, use, and disposal are considered. The idea is to increase growth without increasing the amount of consumable resources. Some considerations include recovery and reuse, disassembly, waste minimization, energy use, material use, and environmental accident prevention. Design for environment techniques may include life cycle assessment, technology assessment, sustainable engineering, and sustainable design.

The design for the environment program of the U.S. Environmental Protection Agency helps consumers, businesses, and institutional buyers identify cleaning and other products that perform well and are safer for human health and the environment. Lists are available that identify safer chemical products for use in manufacturing processes and also aid in determining safer reuse/recovery techniques or appropriate disposal (U.S. Environmental Protection Agency).

Design for Serviceability

Design for serviceability aims to return operation and use easily and quickly after failure. This is often associated with maintainability. Design for serviceability/maintainability objectives include reduction of service requirements and frequency, facilitation of diagnosis, minimization of the time and costs to disassemble, repair/replace and reassembly of the product within the service process, and reduction of the cost of service components.

Design for Ergonomics

Human factors engineering must ensure that the product is designed for the human user. Some of the attributes that may be considered are fitting the product to the user's attributes, simplifying the user's tasks (user-friendliness), making controls and functions obvious and easy to use, and anticipation of human error.

Design specification best practices include (a) use of global anthropometry considerations (North America, Europe, Asia, and Latin America); (b) use of dimensions and ranges that support adjustability and reconfiguration; (c) designs to accommodate neutral postures and task variation; (d) minimization of manual material handling requirements; and (e) environmental considerations such as lighting, temperature, noise and vibration.

Design for Aesthetics

Aesthetics is the human perception of beauty, including sight, sound, smell, touch, taste, and movement. Aesthetics is the aspect of design and technology that most closely relates to art and design, and issues of color, shape, texture, contrast, form, balance, cultural references, and emotional response are common to both areas. Products are becoming smaller and lighter. Customers desire that products be appealing in appearance. While considering customer image requirements, it is important to consider how the product will be manufactured to meet these aesthetic characteristics during the design process. Incorporation of a good industrial design leads to products that are genuinely appealing and represent a synthesis of form and function.

Design for Packaging

The most effective packaging for the product must be determined. The size and physical characteristics of the product are important. Design for automatic packaging methods may be considered. Packaging may be designed for maximum benefit in shipping or for product protection in distribution, storage, sale, and use. Design for packaging incorporates package design and development as an integral part of the new product development process. Products that are designed with packaging in mind can help ensure cost savings and product protection as it moves through the supply chain.

Design for Features

Design for features considers the accessories, options, and attachments that may be used in conjunction with the product. Adding or deleting options is often used in creating products with similar manufacturing characteristics while meeting customer requirements within targeted marketing segments. It allows for expansion of product line offerings without the cost and time involved in complete redesign.

Design for Time to Market

Design for time to market helps ensure that timeliness of product launches may be maintained even as product life cycles continue to shorten. The ability to make the product either earlier or faster than the competition can provide a market leadership advantage. Reducing time to market has a significant positive impact on revenue realization. With optimized processes, prelaunch development costs can be lower, time to launch can be faster, and market share gains can be faster and larger. An example of this is Toyota and their Prius automobile. By being first to market a hybrid gas/electric automobile, Toyota was in a position to establish market leadership in the hybrid market segment through first mover advantage (see Liker, 2004). First mover advantage is the advantage gained by the initial significant occupant of a market segment. Part of this advantage may be attributable to the fact that the first entrant can develop or gain control of resources that followers may not be able to match.

EXAMPLES

See the section of this chapter entitled, "How to Use the Tool" for examples related to how to use this tool.

SOFTWARE

No specific software brand is suggested by the author. There are a number of software programs and vendors that provide software package to support Design for X.

REFERENCES

Gattenby, D.A. and Foo, G. Design for X (DFX): Key to competitive, profitable products. *AT&T Technical Journal*, May–June 1990.

Huang, G.Q. *Design for X, Concurrent Engineering Imperatives*. London: Chapman & Hall, 1996.

Liker, J. *The Toyota Way: 14 Management Principles from the World's Greatest Manufacturer.* New York: McGraw-Hill, 2004.

U.S. Environmental Protection Agency, http://www.epa.gov/dfe/.

SUGGESTED ADDITIONAL READING

Boothroyd, G., Dewhurts, P., and Knight, W. *Product Design for Manufacture and Assembly*. Boca Raton, FL: CRC Press, 2010.

Bralla, J. *Design for Manufacturability Handbook*. New York: McGraw-Hill Professional, 1998.

8

Flowcharting

H. James Harrington

CONTENTS

DEFINITION

Flowcharting is a method of graphically describing an existing or proposed process by using simple symbols, lines, and words to pictorially display the sequence of activities. Flowcharts are used to understand, analyze, and communicate the activities that make up major processes throughout an organization. It can be used to graphically display movement of product, communications, and knowledge related to anything that takes an input and value to it and produces an output.

USER

This tool can be effectively used by an individual to document an activity or process. It can be equally effective when used by a team to gain common understanding or to document an activity, process, or system.

OFTEN USED IN THE FOLLOWING PHASES OF THE INNOVATIVE PROCESS

The following are the seven phases of the innovative cycle. An X after the phase name indicates that the tool/methodology is used during that specific phase.

- Creation phase X
- Value proposition phase X
- Resourcing phase X
- Documentation phase X
- Production phase X
- Sales/delivery phase
- Performance analysis phase

TOOL ACTIVITY BY PHASE

- Creative phase—During the creative stage, flowcharting is frequently used to identify and characterize the present process or system. It is also used to study a proposed solution to define weaknesses, improve on them, and document the final solution.

- Value proposition phase—During the value proposition phase, flow-charting is frequently used to analyze and understand the proposed process. It is then used to calculate key parameters like cost, cycle time, and risk.
- Resourcing phase—During this phase, the flowchart is an effective tool for doing activity-based costing and costs–benefits analysis.
- Documentation phase—During this phase, flowcharting is often used as a key part of the operating instructions and workflow diagrams.
- Production phase—During the production phase, flowcharts are frequently used to communicate the sequence of operations that are required to produce the end output and to communicate key information to the individuals performing the activities within a complex process.

HOW TO USE THE TOOL

Flowcharting is a method of graphically describing an existing or proposed new process by using simple symbols, lines, and words to pictorially display the sequence of activities. Flowcharts are used to understand, analyze, and communicate the activities that make up major processes throughout an organization. They are essential tools used in process redesign, process reengineering, Six Sigma, and ISO 9000 documentation.

General

There are many different types of flowcharts. It is important that the practitioner select the best one for the specific application (see Table 8.1).

Preparation

Flowcharts have application in almost all parts of problem-solving and innovative processes. They are useful for identifying problems, defining measurement points for data collection, idea generation, and idea selection. The usefulness of flowcharts has been recognized to such an extent that structured flowcharts are increasingly used as the basis for computer-aided software engineering (Harrington, 2012).

TABLE 8.1

Seven Types of Flowcharts

Type	Description
1. Process blocks (block diagram):	Document "what" is done, to illustrate a high-level flow of operations.
2. Process charts:	Document "how," by breaking down a process under study into activities chronologically.
3. Procedure charts:	Document the detailed flow of activities in the process.
4. Functional flowcharts:	Document the process, emphasizing responsibilities and interaction between departments.
5. Geographical flowcharts:	Document the physical movement of people and/or materials.
6. Paperwork flowcharts:	Document the detailed flow of paperwork forms within a process.
7. Information flowcharts:	Document office procedures (manual and automated) using standard IS017BM symbols.

Seven types of flowcharts will be discussed below. The *standard process flowchart* is the most widely used, probably because it has the broadest applicability to problem solving and because it can be used as a baseline to create some of the other flowcharts that may better graphically represent a particular point. You should modify your use of flowcharts to your specific needs. Be creative.

An important part of flowcharts is the symbols used to represent various kinds of activities. The American National Standards Institute (ANSI) has developed a standard set of symbols for flowcharting. We have seen many modifications of the ANSI symbol set used effectively in organizations. The important thing is that within an organization, there is consistency among documents.

Standard Flowchart Symbols

The most effective flowcharts use only widely known standard symbols. Think about how much easier it is to read a road map when you are familiar with the meaning of each symbol and what a nuisance it is to have some strange, unfamiliar shape in the area of the map you are using to make a decision about your travel plans (Harrington, 2012).

The flowchart is one of the oldest of all the design aids available. For simplicity, we will review only 10 of the most common symbols, most of which are published by ANSI.

Operation: rectangle. Use this symbol whenever a change in an item occurs. The change may result from the expenditure of labor, a machine activity, or a combination of both. It is used to denote activity of any kind, from drilling a hole to computer data processing. It is the correct symbol to use when no other one is appropriate. Normally, you should include a short description of the activity in the rectangle.

Movement/transportation: fat arrow. Use a fat arrow to indicate movement of the output between locations (e.g., sending parts to stock, mailing a letter).

Decision point: diamond. Put a diamond at the point in the process at which a decision must be made. The next series of activities will vary based on this decision. For example, "If the letter is correct, it will be signed. If it is incorrect, it will be retyped." Typically, the outputs from the diamond are marked with the options (e.g., YES–NO, TRUE–FALSE).

Inspection: square. Use to signify that the process flow has stopped so the quality of the output can be evaluated. It typically involves an inspection conducted by someone other than the person who performed the previous activity. It also can represent the point at which an approval signature is required.

Paper documents: wiggle-bottomed rectangle. Use this symbol to show when the output from an activity included information recorded on paper (e.g., written reports, letters, or computer printouts).

Meanings

Start/end: circle. Use this symbol at the beginning and end points of a flowchart.

Delay: blunted rectangle. Use this symbol, sometimes called a bullet, when an item or person must wait, or when an item is placed in temporary storage before the next scheduled activity is performed (e.g., waiting for an airplane, waiting for a signature).

Storage: triangle. Use a triangle when a controlled storage condition and an order or requisition are required to remove the item for the next scheduled activity. This symbol is used most often to show that output is in storage waiting for a customer. The object of a continuous-flow process is to eliminate all the triangles and blunt rectangles from the process flowchart. In a business process, the triangle would be used to show the status of a purchase requisition being held by purchasing, waiting for finance to verify that the item was in the approved budget.

Annotation: open rectangle. Use an open rectangle connected to the flowchart by a dotted line to record additional information about the symbol to which it is connected. For example, in a complex flowchart plotted on many sheets of paper, this symbol could be connected to a small circle to provide the page number where the inputs will reenter the process. Another way to use an open rectangle is to identify who is responsible for performing an activity or the document that controls the activity. The open rectangle is connected to the flowchart with a dotted line so that it will not be confused with a line arrow that denotes activity flow.

Direction flow: arrow. Use an arrow to denote the direction and order of process steps. An arrow is used for movement from one symbol to another. The arrow denotes direction—up, down, or sideways. The ANSI indicates that the arrowhead is not necessary when the direction flow is from top to bottom or from left to right. However, to avoid misinterpretation by others who may not be as familiar with flowchart symbols, it is recommended that you always use arrowheads.

The 10 symbols listed are not meant to be a complete list of flowchart symbols, but they are the minimum you will need to adequately flowchart your business process. As you learn more about flowcharting, you can expand the number of symbols you use to cover your specific field and needs.

Nine Steps to Prepare a Flowchart

There is really no prescribed sequence for generating flowcharts. However, some common-sense rules should apply in flowcharting. The following general steps have proven to be effective in most flowcharting activities:

1. Before beginning, define your objectives for flowcharting.
2. Determine the process boundaries.
   a. What is included in the process?
   b. What is not included?

     c. What are the outputs from the process?

     d. What are the inputs to the process?

     e. What departments are involved in the process?

3. Select the most appropriate flowchart type(s) for your objectives.
4. Prepare a high-level block diagram of the process you wish to flowchart. Stretch the boundaries of the process at this point to get the broadest view of the process possible. Prepare a block diagram regardless of if you will ultimately use block diagrams or different flowchart types.
5. Determine what people or functional areas are involved in the process. Assemble the most appropriate team to flowchart the process.
6. Determine what tools or standard formats will be used, if this has not already been established by a broader, perhaps company-wide, initiative.
7. Begin flowcharting the process. Start at a high level and work toward more detail. Pay close attention to all interactions among people, departments, and functions. Using the guidelines and tips below, identify the suppliers of all process inputs and the customers of all process outputs.
8. Early in the creation of flowcharts, review the appropriateness of the process boundaries. Modify the process definition, based on its boundaries, as necessary. Also, modify the team members, if necessary, to reflect any changes in process definition.
9. Complete the flowcharts.

Guidelines and Tips

The level of detail of your flowcharts is largely a matter of judgment, and this will improve with experience. By fully understanding your objectives before starting to flowchart, you will have a better idea of the meaningful level of detail. Early in the flowcharting activity, it is often a good idea to take a small part of the process and dissect it with the team. Take it to a very fine level of detail and then agree with the team on the appropriate level of detail, based on the team's objectives.

There are two fundamentally different ways to approach flowcharting. One way is to start at the beginning and define each step in great detail. Another way is to build a simple view of the process first and then add detail. While you should ultimately choose the method you are most comfortable with, our experience has been that it is generally easier to start simple and add detail.

It is very important to separate the flowcharting process from process improvements. Focus first on completing flowcharting. Create an *issues list* and defer judgment on this list until flowcharting is complete. Discussions on potential improvements can be a great distraction and delay the overall improvement process.

It is important to flowchart processes as they are today or to flowchart a vision of the future. It is crucial not to mix these in the same flowchart. Effective flowcharting, therefore, may require interviewing skills to uncover the actual processes used today. Flowcharting may also be supplemented with process walk-throughs to ensure that current processes are accurately captured. Walk-throughs are a highly recommended part of the process. You will be surprised how many things you find during a walk-through that are different from the way you have documented the process. The steps to a process walk-through are as follows:

1. Identify the scope of the process to be reviewed.
2. Develop the objectives of the walk-through.
3. Determine the walk-through method:
   a. Interview
   b. Observation
   c. Sampling
4. Create the interview worksheet (who, what, when).
5. Conduct the walk-through.
6. Update the flowchart.

Some of the discoveries of process walk-throughs are

- Differences between the documented process and present practice
- Differences among employees in the way they perform the activity
- Employees in need of retraining
- Suggested improvements to the process, identified by the people performing the process
- Activities that need to be documented
- Process problems, such as
  - Duplication
  - Rework
  - Waste
  - Bureaucracy
- Roadblocks to process improvement
- New training programs required to support the present process

Put dates on all flowcharts created. Flowcharting is an iterative process. Including dates on all revisions will save you many headaches later on.

Value Stream Mapping

Value stream mapping leaves a lot of latitude to the PIT (process improvement team). There are many different symbols that can be used, and the team is allowed to add additional symbols to meet their particular needs (see below and Figure 8.1). Some of the more common symbols are the following:

- A truck—to show that the movement is by truck
- A train—to show that the movement is by train

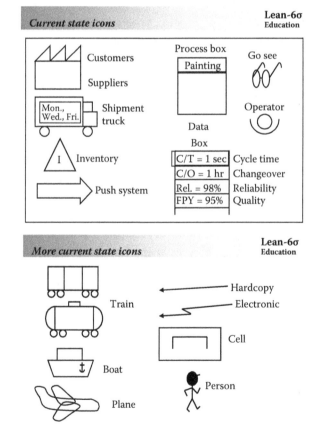

FIGURE 8.1
Value stream mapping symbols (icons).

- An airplane—to show that the movement is by airplane
- A stick figure—to show that the movement is by people carrying it from one spot to another
- A triangle with a q inside—to show that the product is in queue or waiting
- A triangle with a capital I inside—to show that it is in inventory
- A starburst—to show that an activity has some improvement activities assigned to it
- An elongated rectangle with FIFO—to show that first in, first out is being practiced
- A pair of eyeglasses—to indicate that the team should look at this particular area
- A thin line with an arrow—to show it is a hard copy that is being transmitted
- An irregular thin line with an arrow—to show that the information is being sent electronically

See Figure 8.2 for a future state value stream map.

Future state map

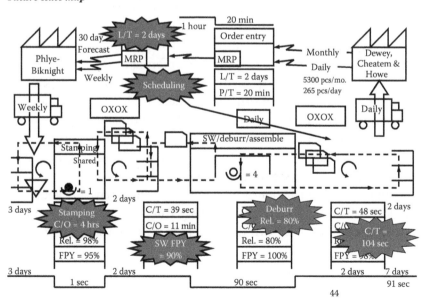

FIGURE 8.2
Future state value stream map.

Graphic Flowcharts

A graphic or physical layout flowchart analyzes the physical flow of activities. It helps minimize the time wasted while work outputs are moved between activities.

Graphics flowcharting is a useful tool for evaluating department layout and paper work flow, and for analyzing product flow by identifying excessive travel and storage delays. In streamlined process improvement (SPI), geographic flowcharting helps in analyzing traffic patterns around busy areas like file cabinets, computers, and copiers. Figure 8.3 is a graphic flowchart of a new employee's first day at XYZ Company.

In addition to the basic flowcharting we have discussed, there is also information diagramming, often with its own set of symbols. As a rule, these are more interesting to computer programmers and automated systems analysts than to managers and employees charting business activities; however, they are important in designing an organizational structure. You make a set of these types of flowcharts that follow information through the process. As you prepare flowcharts, think of your organization's activities in terms of information processing, beginning with your organization's files. They are valuable because they contain information that is created or used by your business processes. Figure 8.4 is a block diagram with its information/communication support system added to it with broken lines.

Next, consider your employees. You and your co-workers have skills of various levels and types. Obviously, even a single worker's knowledge is substantially more sophisticated than the information in the file, but the principle still holds. The employee's value to an organization depends on his or her contribution of information. Whether it is how to load a pallet, introduce a new product, or resolve conflict, information is a resource. This is particularly true in the service industry, which in the 21st century employs more than 80% of the US labor force; all of them can be considered information processors and providers.

If you take an information-processing view when preparing your flowcharts, it will create a common focus on getting and using quality inputs to produce quality outputs. At the same time, this type of view helps people determine who their interfaces are.

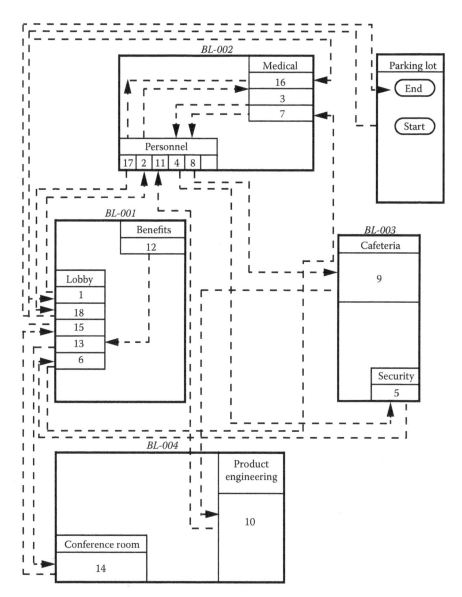

FIGURE 8.3
Graphic flowchart of a new employee at XYZ Company.

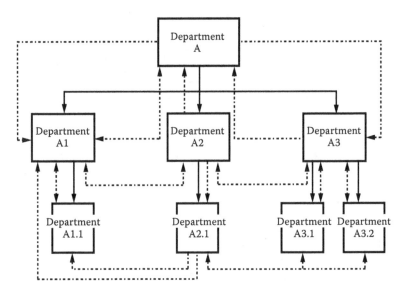

FIGURE 8.4
Typical information/communication flowchart.

Process Knowledge Mapping

There will be an increased emphasis over the next few years on taxonomies, ontologies, and knowledge.

French Caldwell
Vice President, Information and Knowledge Management,
Gartner Group

A process-based knowledge map is a map or diagram that visually displays knowledge within the context of the business process. It shows how knowledge should be used within the process and the source of this knowledge. Any type of knowledge that drives the process or results from execution of this process can be and should be met. This includes tacit (soft) knowledge or explicit (hard) knowledge. Tactic knowledge is defined as undocumented, intangible factors embedded in the individual's experience. Explicit knowledge, on the other hand, is documented and quantified knowledge that is recorded and can be easily distributed.

As you look at your process-based knowledge map, ask the following questions:

- What knowledge is missing?
- What knowledge is most valuable?

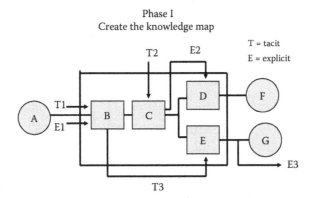

Phase I
Create the knowledge map

FIGURE 8.5
Typical knowledge map.

- What knowledge can be used in other processes?
- What knowledge is generated but not shared?

Knowledge management is often abstract, overly strategized, and weak in implementation. Process-based knowledge maps are concrete and tactical; they allow the business to really focus in on its key knowledge assets. Figure 8.5 is a typical knowledge map.

Other Process Mapping Tools

We have only highlighted the major process mapping tools that we like to use. There are many other tools that are available to the PIT. Some of the others include

- Organizational structure diagrams
- Administrative business processes principles diagrams
- Hierarchical overview diagram
- Global overview of processes and divisions diagrams
- Global process diagram
- Detailed process diagrams
- Form management diagrams
- Form circulation diagrams
- Accounting system diagrams
- Etc.

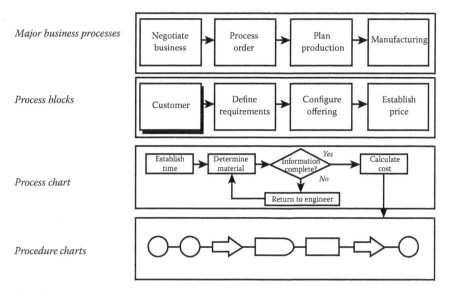

FIGURE 8.6
Relationships among flowcharting techniques.

For information on these diagrams and others, see the book *Business Process Improvement Workbook* by Esseling et al. (1997).

EXAMPLE

See examples in Figures 8.6 through 8.12.

Figure 8.12 shows a detailed flowchart for the 8 activities and 35 tasks of a typical business case process.

SOFTWARE

There are currently a number of software products on the market that can assist you in creating process flowcharts. Some of them are generic tools such as drawing programs, while others have been created specifically for flowcharting. One product we suggest is a software package produced by Edge Software Inc. called WorkDraw. This is an advanced process modeling application that includes flowcharting capabilities.

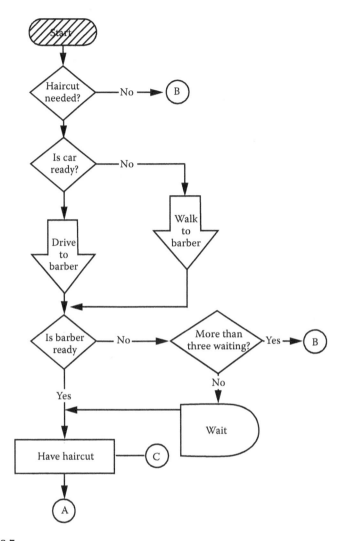

FIGURE 8.7
Two standard process flowcharts of different parts of the process of getting a haircut and/or going fishing (part I).

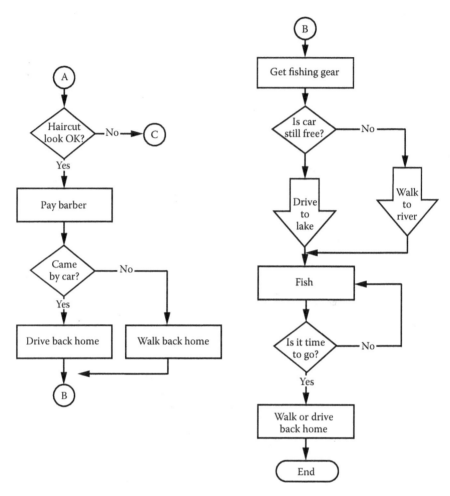

FIGURE 8.8

Two standard process flowcharts of different parts of the process of getting a haircut and/
or going fishing (part II).

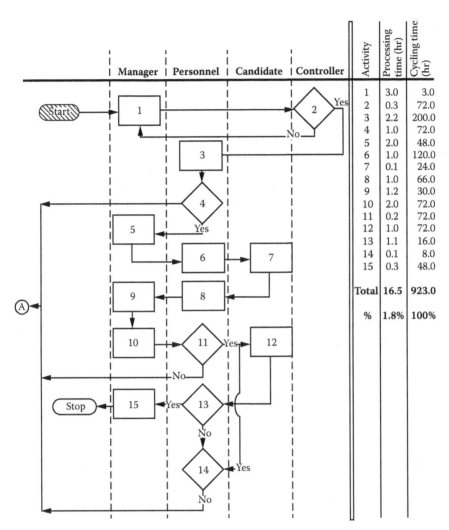

The table within the figure:

Activity	Processing time (hr)	Cycling time (hr)
1	3.0	3.0
2	0.3	72.0
3	2.2	200.0
4	1.0	72.0
5	2.0	48.0
6	1.0	120.0
7	0.1	24.0
8	1.0	66.0
9	1.2	30.0
10	2.0	72.0
11	0.2	72.0
12	1.0	72.0
13	1.1	16.0
14	0.1	8.0
15	0.3	48.0
Total	16.5	923.0
%	1.8%	100%

FIGURE 8.9
Functional flowchart of an internal job search process.

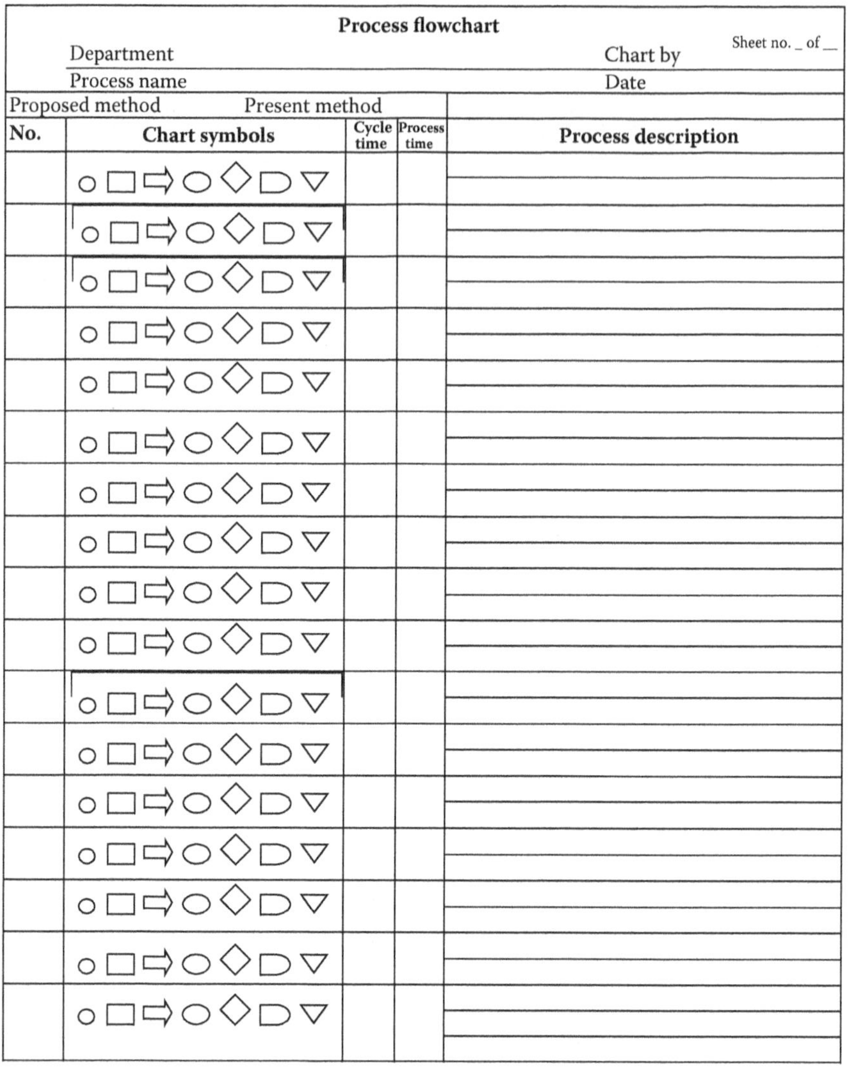

	Process flowchart			Sheet no. _ of _
	Department		Chart by	
	Process name		Date	
Proposed method	Present method			

No.	Chart symbols	Cycle time	Process time	Process description
	○ ▢ ⇨ ○ ◇ ▷ ▽			
	○ ▢ ⇨ ○ ◇ ▷ ▽			
	○ ▢ ⇨ ○ ◇ ▷ ▽			
	○ ▢ ⇨ ○ ◇ ▷ ▽			
	○ ▢ ⇨ ○ ◇ ▷ ▽			
	○ ▢ ⇨ ○ ◇ ▷ ▽			
	○ ▢ ⇨ ○ ◇ ▷ ▽			
	○ ▢ ⇨ ○ ◇ ▷ ▽			
	○ ▢ ⇨ ○ ◇ ▷ ▽			
	○ ▢ ⇨ ○ ◇ ▷ ▽			
	○ ▢ ⇨ ○ ◇ ▷ ▽			
	○ ▢ ⇨ ○ ◇ ▷ ▽			
	○ ▢ ⇨ ○ ◇ ▷ ▽			
	○ ▢ ⇨ ○ ◇ ▷ ▽			
	○ ▢ ⇨ ○ ◇ ▷ ▽			
	○ ▢ ⇨ ○ ◇ ▷ ▽			
	○ ▢ ⇨ ○ ◇ ▷ ▽			

FIGURE 8.10
Sample flowchart.

Process flowchart				
Department 300				Sheet No. 1_ of 1__
				Chart by HJH
Process name Internal job research process				Date Jan 16, 2012
Proposed method Present method				

No.	Chart symbols	Cycle time	Process time	Process description
1	○ □ ⇨ ○ ◇ ▷ ▽			Recognize need. Complete payback analysis. Prepare personnel requisition. Prepare budget request.
2	○ □ ⇨ ○ ◇ ▷ ▽			Evaluate budget. If yes, sign personnel recognition slip. If no, return total package with reject letter to manager.
3	○ □ ⇨ ○ ◇ ▷ ▽			Conduct in-house search.
4	○ □ ⇨ ○ ◇ ▷ ▽			If in-house candidates exist, provide list to management. If not, start outside hiring procedure.
5	○ □ ⇨ ○ ◇ ▷ ▽			Review candidates' paperwork and prepare a list of candidates to be interviewed.
6	○ □ ⇨ ○ ◇ ▷ ▽			Have candidates' managers review job with the employees and determine those who are interested in the position.
7	○ □ ⇨ ○ ◇ ▷ ▽			Notify personnel of candidates interested in being interviewed.
8	○ □ ⇨ ○ ◇ ▷ ▽			Set up meeting between manager and candidates.
9	○ □ ⇨ ○ ◇ ▷ ▽			Interview candidates and review details of job.
10	○ □ ⇨ ○ ◇ ▷ ▽			Notify personnel of interview results.
11	○ □ ⇨ ○ ◇ ▷ ▽			If acceptable candidate is available, make job offer. If not, start outside hiring process.
12	○ □ ⇨ ○ ◇ ▷ ▽			Evaluate job offer and notify personnel of candidates decision.
13	○ □ ⇨ ○ ◇ ▷ ▽			If yes, notify manager that the job has been filled. If no, go to activity 14.
14	○ □ ⇨ ○ ◇ ▷ ▽			Were there other acceptable candidates? If yes, go to activity 12. If no, start outside hiring process.
15	○ □ ⇨ ○ ◇ ▷ ▽			Have new manager contact candidate's present manager and arrange for the candidate to report to work.
16	○ □ ⇨ ○ ◇ ▷ ▽			Start outside hiring process.

O = Start or stop the process ● = Inspect ⇨ = Movement ▭ = Operation ◇ = Decision point ▷ = Delay ▽ = Storage

FIGURE 8.11

Flowchart of an internal job research process.

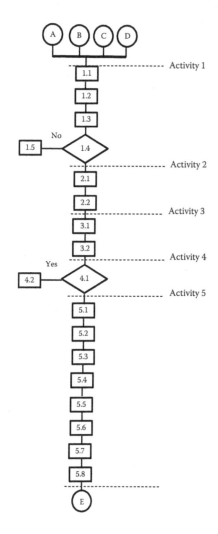

A. Input—Value propositions
B. Input—Research evaluations
C. Input—Proposals without value propositions
D. Input—Business case preparation activities' outputs

Activity 1—Set the proposal context and stimulus

- Task 1.1 Create the BCD team
- Task 1.2 Preparing the BCD team
- Task 1.3 Analysis of proposed project's input documents
- Task 1.4 Does the proposal meet the required ground rules to prepare a business case?
- Task 1.5 If the answer to Task 1.4 is no, then take appropriate action

Activity 2—Define the sponsor's role and test alignment to organizational objectives

- Task 2.1 Define the business case sponsor's role in the BCD process
- Task 2.2 Align the project/initiative with strategic goals and objectives

Activity 3—Prepare the BCD team's charter and output

- Task 3.1 Develop the BCD team's charter
- Task 3.2 Define the business case final report

Activity 4—Patent, trademark, copyright considerations

- Task 4.1 Is the idea/concept an original idea/concept?
- Task 4.2 Start patent/copyright process

Activity 5—Collect relevant information/data

- Task 5.1 Characterizing the current state
- Task 5.2 Characterizing proposed future state
- Task 5.3 Define the proposed future state assumptions
- Task 5.4 Define the implementation process
- Task 5.5 Define the major parameters related to the proposal
- Task 5.6 Define the quality and type of data to be collected and prioritized
- Task 5.7 Develop the data collection plan
- Task 5.8 Collecting process/product installation–related data or information

FIGURE 8.12

Detailed flowchart of typical business case. (*Continued*)

- Visio 14.0.6
- SmartDraw VP 19.1.3.2
- Flowcharter 14.1.2
- Edraw Flowchart 6
- EDGE Diagrammer 6.24
- WizFlow Flowcharter 6.24
- RFFlow 5.06

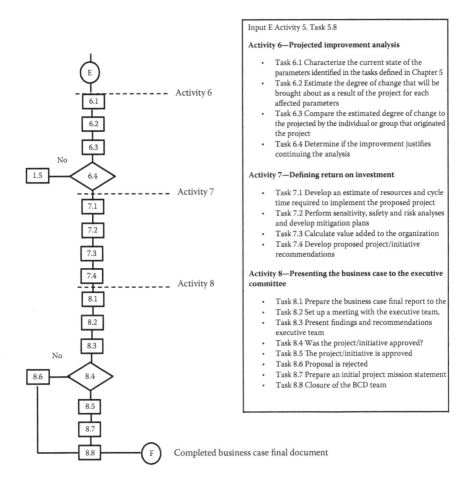

Input E Activity 5. Task 5.8

Activity 6—Projected improvement analysis

- Task 6.1 Characterize the current state of the parameters identified in the tasks defined in Chapter 5
- Task 6.2 Estimate the degree of change that will be brought about as a result of the project for each affected parameters
- Task 6.3 Compare the estimated degree of change to the projected by the individual or group that originated the project
- Task 6.4 Determine if the improvement justifies continuing the analysis

Activity 7—Defining return on investment

- Task 7.1 Develop an estimate of resources and cycle time required to implement the proposed project
- Task 7.2 Perform sensitivity, safety and risk analyses and develop mitigation plans
- Task 7.3 Calculate value added to the organization
- Task 7.4 Develop proposed project/initiative recommendations

Activity 8—Presenting the business case to the executive committee

- Task 8.1 Prepare the business case final report to the
- Task 8.2 Set up a meeting with the executive team,
- Task 8.3 Present findings and recommendations executive team
- Task 8.4 Was the project/initiative approved?
- Task 8.5 The project/initiative is approved
- Task 8.6 Proposal is rejected
- Task 8.7 Prepare an initial project mission statement
- Task 8.8 Closure of the BCD team

Completed business case final document

FIGURE 8.12 (CONTINUED)
Detailed flowchart of typical business case.

REFERENCES

Esseling, E.K.C., Harrington, H.J., and VanNimwegen, H. *Business Process Improvement Workbook.* New York: McGraw-Hill, 1997.

Harrington, H.J. *Streamlined Process Improvement.* New York: McGraw-Hill, 2012.

SUGGESTED ADDITIONAL READING

Galloway, D. *Mapping Work Processes.* Milwaukee, WI: ASQ Quality Press, 1994.

Harrington, H.J. *Business Process Improvement*. New York: McGraw-Hill, 1991.

Harrington, H.J. *The Business Process Improvement Workbook*. New York: McGraw-Hill, 1997.

Harrington, H.J., Hoffherr, G., and Reid, R. *Area Activity Analysis*. New York: McGraw-Hill, 1998.

PQ Systems Inc. *Total Quality Tools*. Milwaukee, WI: ASQ Quality Press, 1996.

9

Force Field Analysis

H. James Harrington

CONTENTS

There is a good and bad in all of us that we wrestle with all the time. The same is true in the processes we design.

H. James Harrington

DEFINITION

Force field analysis is a visual aid for pinpointing and analyzing elements that resist change (restraining forces) or push for change (driving forces). This technique helps drive improvement by developing plans to overcome the restrainers and make maximum use of the driving forces.

USER

This tool is most effectively used by a group of people representing different parts of the business who are familiar with the item being discussed.

OFTEN USED IN THE FOLLOWING PHASES OF THE INNOVATIVE PROCESS

The following are the seven phases of the innovative cycle. An X after the phase name indicates that the tool/methodology is used during that specific phase.

- Creation phase X
- Value proposition phase X
- Resourcing phase
- Documentation phase
- Production phase X
- Sales/delivery phase X
- Performance analysis phase

TOOL ACTIVITY BY PHASE

- Creation phase—During this phase, force field analysis can be used to identify opposing forces that offset each other. This analysis is used to develop an innovative concept that is designed to offset the negative impacts as much as possible.
- Value proposition phase—During this phase, force field analysis is used to show the negative and positive impacts of the proposed innovative concept.
- Production phase—During this phase, force field analysis can be used to analyze the advantages and disadvantages to various production and procurement opportunities.
- Sales/delivery phase—Force field analysis approaches can be used to weigh the pros and cons of the various potential marketing, advertising, and sales approaches.

HOW THE TOOL IS USED

Overview

The force field analysis technique has been used in a number of settings to do the following:

- Analyze a problem situation into its basic components.
- Identify those key elements of the problem situation about which something can realistically be done.
- Develop a systematic and insightful strategy for problem solving that minimizes *boomerang* effects and irrelevant efforts.
- Create a guiding set of criteria for the evaluation of action step.

The technique is an effective device for achieving each of these purposes when it is seriously employed.

Kurt Lewin, who developed force field analysis, has proposed that any problem situation, be it the behavior of an individual or group, the current state or condition of an organization, a particular set of attitudes, or frame of mind—may be thought of as constituting a level of activity that is somehow different from that desired. For example, smoking, as an activity, may become the basis for a problem when it occurs with greater intensity or at a higher level than one may desire. Quality, as another example of an activity level, may become a problem when it is at a lower-than-desirable level. Depression or authoritarianism, as an example of attitudinal activity level, can become a problem when it is too intense or at a higher-than-desirable level.

The level of the activity, to put it differently, is the starting point in problem identification and analysis. To constitute a problem, the current level typically departs from some implicit norm or goal.

A particular activity level may be thought of as resulting from a number of pressures and influences acting on the individual, group, or organization in question. Lewin calls these numerous influences *forces*, and they may be both external to and internal to the person or situation in question. Lewin identifies two kinds of forces:

- Driving or facilitating forces that promote the occurrence of the particular activity of concern
- Restraining or inhibiting forces that inhibit or oppose the occurrence of the same activity

An activity level is the result of the simultaneous operation of both facilitating and inhibiting forces. The two force fields push in opposite directions and, while the stronger of the two will tend to characterize the problem situation, a point of balance is usually achieved, which gives the appearance of habitual behavior or of a steady-state condition. Changes in the strength of either of the fields, however, can cause a change in the activity level of concern. Thus, apparently habitual ways of behaving, or frozen attitudes, can be changed (and related problems solved) by bringing about changes in the relative strengths of facilitating and inhibiting force fields.

To appreciate just what kinds of forces are operating in a given situation and which ones are susceptible to influence, a force field analysis must be made. As a first step to a fuller understanding of the problem, the forces (both facilitating and inhibiting) should be identified as fully as possible. Identified forces should be listed and, as much as possible, their relative contributions or strengths should be noted.

Basic Steps in Force Field Analysis

Once the problem has been recognized, and commitment is made by the appropriate stakeholders to change the problem situation, there are four basic steps used in the force field analysis activity to analyze the problem (Harrington and Lomax, 2009):

1. Define the problem and propose an ideal solution.
2. Identify and evaluate the forces acting on the problem situation.
3. Develop and implement a strategy for changing these forces.
4. Reexamine the situation to determine the effectiveness of the change and make further adjustments if necessary.

Let us examine these steps.

Step One

The first step includes defining the problem and proposing the ideal situation. In defining the problem, it is necessary to say exactly what it is:

1. Propose an ideal situation in a *goal statement*. It can be prepared by answering the question, "What will the situation be like when

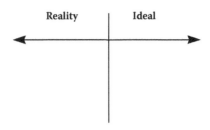

FIGURE 9.1
Reality versus ideal.

the problem is solved?" The answer must be tested to determine if it really gets to the heart of the problem.

2. Another possible question is, "What would the situation be like if everything were operating ideally?" (see Figure 9.1).

Determining the precise goal statement is important because it guides the rest of the problem-solving steps.

Step Two

The second step is to identify and evaluate the forces that act on the problem. The goal of this step is to identify restraining or inhibiting forces.

1. Facilitating forces tend to move the problem situation from reality toward the ideal. Restraining forces resist the movement toward the ideal state, and, in a state of equilibrium, counterbalance the facilitating forces (see Figure 9.2).
2. You can visualize a problem situation by drawing a line down a sheet of paper and listing the facilitating forces on one side and the restraining

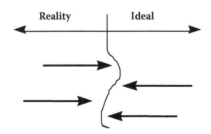

FIGURE 9.2
As-is state in equilibrium.

Facilitators	Restrainers

FIGURE 9.3
In balance.

forces on the other side. Each of these forces has its own weight and, taken together, they keep the field in balance (see Figure 9.3).

In addition to helping make the problem situation visual, force field analysis provides a method for developing a solution. The most effective solution will involve reducing the restraining forces operating on the problem (see Figure 9.4).

3. There are two reasons for reducing the restraining forces:
   • To move the problem toward the solution
   • To avoid the effect of having too many facilitating forces

   Because the forces on each side of the situation are in balance, removing or reducing the restraining forces will cause movement of the problem toward the solution.

4. On the other hand, adding facilitating forces without reducing restraining forces will likely lead to the appearance of new restrainers. Remember that although you may change the situation by changing a force, you may not have improved the situation.

5. An effective strategy cannot be planned without evaluating the restraining forces for two factors: first, whether and to what degree a restrainer is changeable, and secondly, to what degree will changing a restrainer affect the problem.

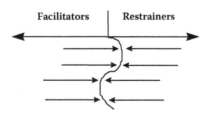

FIGURE 9.4
Reducing the restraining forces.

6. It is ineffective and a waste of energy to try to change an unchangeable force.

7. One way to begin planning a strategy is to evaluate each force to see how changeable it is. A simple three-point rating scale is sufficient.
   - A fixed, unchanging force
     - Example: a contractual item, a law, a fixed budget
   - A force changeable with moderate to extensive effort
     - Example: an item that involves the efforts and cooperation of many departments
   - A change that can be rather readily performed, perhaps by just revising a procedure and is probably within the control of the group

8. The change or removal of some restrainers may have little or no impact. You must consider the effect that changing the force will have. It is good, then, to also rate the restrainers for their effect on solving the problem.

9. A three-point rating scale can be used to rate the effect a change will have on the problem:
   - No significant improvement will occur with the change.
   - Some minor improvement will occur with the change—perhaps up to 20% of the improvement needed to solve the problem.
   - A major improvement results from changing this force, that is, from 25% to 100% of the needed improvement.

10. After you have rated all of the forces operating on the problem situation, you can determine a priority for dealing with each force by adding together the numbers with which you rated each of the forces. The highest priority will be the restraining force, which will have the most effect and which is most changeable. Next will be those forces that you judge to have a large effect, but are less changeable, and so on.

Step Three

At this point in the force field analysis, you are ready to begin the third step: developing and implementing a strategy for changing the forces effecting a situation.

1. In deciding the priority, strive for a balance of ease of change and the impact of the proposed change. Often the actions in dealing with any situation will require creative thinking. The balance between

the facilitators and the restrainers is a clue for deciding which forces to change.

2. A recommended tool is to remove the restraining forces to allow the point of equilibrium to shift. If the new point is not satisfactory, examine the driving forces and determine which ones you can successfully change.

Step Four

The fourth step is to examine the situation. If you are still not satisfied with the new situation, determine which facilitating forces can be added.

Each time a change is planned, take the time to estimate and determine whether the change was worth it. Ask these questions:

- Will it produce the desired results?
- Which facilitating and restraining factors will be affected and by how much?
- How will the equilibrium point be affected?
- Is there a better way of getting the same results?
- Does the change have a negative impact on other parts of the process?
- What will be the return on investment?

Force field analysis is a straightforward tool. Using it with diligence and an ongoing evaluation of solutions will assure that it can work toward the achievement of your desired goal. Force field analysis is valuable because it goes beyond brainstorming by helping develop plans and set priorities.

EXAMPLE

Consider a *real* example of *starting a TQM*, as shown in Figure 9.5. In this example, the organization would have to consider if the restraining forces might be too great to overcome. If the organization decides to continue with starting a TQM, a great deal of effort must be expended to overcome the restraining forces.

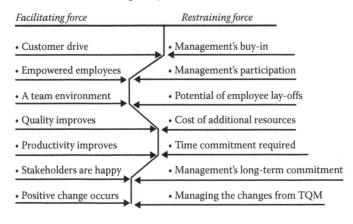

Starting a TQM effort

FIGURE 9.5
Example of a completed force field analysis diagram.

SOFTWARE

Some commercial software available includes but is not limited to

- Edraw max: http://www.edrawsoft.com
- Smartdraw: http://www.smartdraw.com
- Affinity Diagram 2.1: http://mobile.brothersoft.com/
- QI macros: http://www.qimacros.com

REFERENCE

Harrington, H.J. and Lomax, K. *Performance Improvement Methods*. Chico, CA: Patent Professional, 2009.

SUGGESTED ADDITIONAL READING

Harrington, H.J., Gupta, P., and Voehl, F. *The Six Sigma Green Belt Handbook*. Chico, CA: Paton Press, 2009.
Wortman, B. *CSSBB*. West Terre Haute, IN: Quality Council of Indiana, 2001.

10

Kano Model

H. James Harrington

CONTENTS

When you meet customer requirements, you do not gain customer loyalty. You need to WOW them to have customers for life.

H. James Harrington

DEFINITION

The Kano model is a pictorial way to look at customer levels of dissatisfaction and satisfaction to define how they relate to the different product characteristics. The Kano method is based on the idea that features can be plotted using axes of fulfillment and delight. This defines areas of *must haves, more is better,* and *delighters.* It classifies customer preferences into five categories.

1. Attractive
2. One-dimensional
3. Must-be
4. Indifferent
5. Reverse

USER

This tool can be used by individuals, but its best use is with a group of four to eight people. Cross-functional teams usually yield the best results from this activity.

OFTEN USED IN THE FOLLOWING PHASES OF THE INNOVATIVE PROCESS

The following are the seven phases of the innovative cycle. An X after the phase name indicates that the tool/methodology is used during that specific phase.

- Creation phase X
- Value proposition phase X
- Resourcing phase
- Documentation phase
- Production phase
- Sales/delivery phase X
- Performance analysis phase

TOOL ACTIVITY BY PHASE

- Creation phase—The Kano model is used during this phase to define what the customer really requires and what will delight the customer, setting the output from the innovative process apart from the output from other organizations. It provides a key input to the design specification as well as an effective way to measure the potential of any completed design.
- Value proposition phase—During this phase, the Kano model is used to determine the value the external consumer/customer will put on the output from the innovation process.

HOW THE TOOL IS USED

The Kano model looks at the product/output attributes of three categories:

- Threshold (basic or must-be) attributes: These are the attributes that the customer must have. They do not provide an opportunity for product differentiation.
- Performance (one-dimensional) attributes: These are attributes that will increase customer satisfaction. The better or more of these attributes you are offering, the higher your chances of higher levels of customer satisfaction.
- Excitement (attractive) attributes: These are attributes that will *wow* the customer. They are the ones that are unspoken or documented by the customer. They are unexpected by the customer but will result in high levels of customer satisfaction when they are provided. The absence of these attributes does not lead to customer dissatisfaction.

The model provides insight into the attributes of a product or service that the customer perceives to be important. It is used to help develop better understanding of the importance of the different product or service attributes and their impact on the customer. The five classifications are useful in helping define when *good is good enough* and when *more is better*.

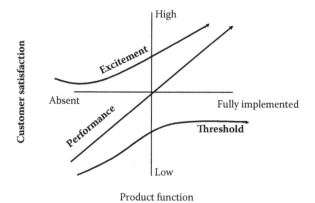

FIGURE 10.1
Kano model.

Often in using the Kano model, the user will focus on three categories: *must-be*, *one-dimensional*, and *excitement*. A product or service must meet the basic attributes at the maximized performance attribute level. A wise supplier includes as many excitement attributes as possible at a cost the customer can afford. It is the excitement attributes that set an organization apart from its competitors. Meeting the *must-be* and performance requirements in most cases does not make the supplier a perfect supplier, but not meeting the *must-be* and performance requirements is very quickly recognized as poor performance (see Figure 10.1) (Kano et al., 1984).

Voice of the Customer

Threshold (Must-Be) Quality Attributes

Threshold (must-be or basic) quality attributes are the expected attributes or *musts* of a product or service and do not provide an opportunity for product differentiation. Increasing the performance of these attributes provides diminishing returns in terms of customer satisfaction; however, the absence of good performance of these attributes results in extreme customer dissatisfaction. An example of a threshold attribute would be brakes on a car; when they do not work, the driver is extremely dissatisfied.

These threshold quality attributes are not typically captured in quality function deployment (QFD) or other evaluation tools as products are not rated on the degree to which a threshold attribute is met; the attribute is either satisfied or not (see Figure 10.2).

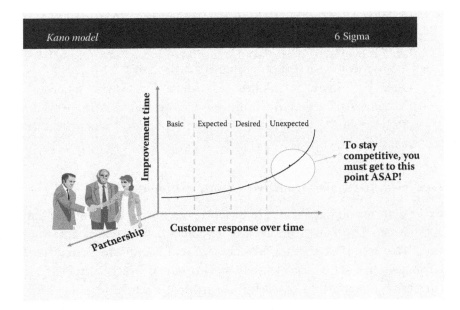

FIGURE 10.2
Customer response level over time.

The following are "Kano threshold quality attributes" rules:

- Must be present or the customer is dissatisfied
- May not raise satisfaction when present

Performance Quality Attributes

Performance quality attributes are those for which more is generally better and will improve customer satisfaction. Conversely, an absent or weak performance attribute reduces customer satisfaction. Of the needs customers verbalize, most will fall into the category of performance attributes. These attributes will form the weighted needs against which product or service concepts will be mentally evaluated by the customer.

The price a customer is willing to pay for a product or service is closely tied to performance quality attributes. For example, some customers would be willing to pay more for a car that provides them with better fuel economy (more is better). In today's fast-paced world, speed is becoming an important attribute because it saves time. No one can have enough time.

The following are characteristics of performance quality attributes:

- More is generally better.
- Poor performance reduces customer satisfaction.
- Often, performance attributes are attributes like faster or quieter.
- Customers are generally willing to pay more; that is, customers generally are willing to pay more for a car that provides them with better fuel economy.

Excitement (Delighter/Attractive) Quality Attributes

Excitement quality attributes are unspoken and unexpected by customers, but can result in high levels of customer satisfaction should they work. However, their absence does not lead to dissatisfaction. Excitement attributes often satisfy latent needs—real needs of which customers are currently unaware. In a competitive marketplace where manufacturers' products provide similar performance and outstanding service is a surprise, providing excitement attributes that address *unknown needs* can provide a competitive advantage. However, they have followed the typical evolution to a performance rather than to a threshold quality attribute. Cup holders were initially excitement attributes that are now *must-be* quality attributes.

The following are characteristics of excitement quality attributes:

- Unspoken and unexpected by customers
- Can result in high levels of customer satisfaction
- Often satisfy latent needs
- Can provide a competitive advantage

Applying Kano Model Analysis

A relatively simple approach to applying Kano model analysis is to ask customers two simple questions for each attribute:

- Rate your satisfaction if the product has this attribute.
- Rate your satisfaction if the product did not have this attribute.

Customers should be asked to answer with one of the following responses:

- Satisfied
- Neutral (it is normally that way)

- Dissatisfied
- Don't care

From these answers, the product or service can be evaluated to define where each attribute falls in the Kano model.

Analysis

Attributes that receive a *neutral* response to question 1 and the *dissatisfied* response to question 2 indicate that exclusion of these attributes in a product or service has the potential to severely influence how the customer will respond. Eliminate or include performance or excitement attributes when their presence or absence may lead to customer dissatisfaction. This often requires a trade-off analysis against cost.

Customers frequently rate attributes or functionality as important. A good question to ask is, "How much extra would you be willing to pay for this attribute or for more of this attribute?" This approach will aid in trade-off decisions, especially for performance attributes.

Consideration should be given to attributes receiving a "don't care" response as they will not increase customer satisfaction or motivate the customer to pay an increased price for the product. However, do not immediately dismiss these attributes if they play a critical role in the product functionality or if they are necessary for other reasons than to satisfy the customer.

All needs cannot be treated equally. When resources are constrained, the focus of the design or improvement project should be to address the most important needs of the most important customers. Prioritizing needs is therefore an important part of the *voice of the customer* (VOC).

Prioritize needs qualitatively as *needs*, *wants*, and *wishes*.

- Needs must be satisfied.
- Wants should be satisfied.
- Wishes can be satisfied.

We suggest using ranking or rating scales (see Figure 10.3).

Understanding Customer Needs

Finding and really understanding how we can best satisfy customers is a challenging and never-ending, task. Because the world is constantly

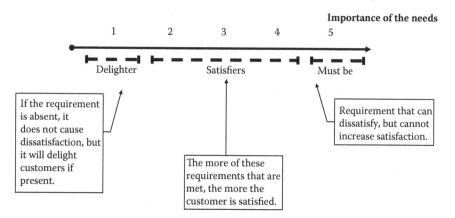

FIGURE 10.3
Customer input rating scale.

changing, so are our customers' expectations. What is great today may not be great tomorrow. For this reason, we must constantly pay attention to our customers. In Baldrige terms, we must establish listening posts from which we can hear how we are doing in our customer's terms. We must understand the competitive environment and the social and cultural environment so we can proactively incorporate improvements to our processes in ways that will meet and exceed customer expectations. If we do this well, our customers will view us as a solution they can count on.

Obtaining Information on Customer Needs

There are many ways to obtain customer information. Some of the most common ways are shown in Table 10.1.

However, a bias should be toward direct communication methods such as focus groups, interviews, and customer observation. These three forms provide the richest and most reliable information. This is because full communication goes beyond words and text.

Surveys, complaints, and even market research can interject bias into the information, hide important facts, or unintentionally misrepresent what is actually true. However, the most robust systems of gathering customer information contain all of these elements in some form. This makes it necessary for testing of data and other validations to be done, looking for consistency across all the data.

TABLE 10.1

Typical Ways to Obtain Customer Information

Method	When Applicable
Interview	To generate broad themes that are of interest and important to customers, particularly if the product or service is new.
Focus group	To obtain more detailed information about a few select topics and to stimulate group interaction.
Survey	To get quantitative information about the importance of needs extracted from an interview or focus group.
Observations	To collect unstated need information by observing how customers interact with the product or service.
Contextual inquire	A way of collecting context-sensitive-needs data by asking questions in the context of their use of the product or service.

VOC data can be collected in two ways:

- Actively, where the team develops a data collection plan and implements the plan
- Passively, where the team receives VOC information from various *listening posts*, such as (see Figure 10.4):
  - Complaints
  - Customer service interactions
  - Online product reviews
  - Returns
  - Trouble ticket data
  - Inputs from sales staff

Listen to Customer Needs

The VOC is often expressed as solutions, complaints, or stated too abstractly to give meaningful information for the process. However, by listening to customer needs (stated in their language) and carefully translating their language into customer requirements (i.e., language of the process, critical-to-quality processes) provides a key benefit. It gives us the opportunity to act on defined customer needs, both current and latent.

This is not rewriting the customer needs; it is merely a translation, like French to English. This translation is essential as much of what is assessed are gaps between process realities, that is, what the process is actually

Active data	Passive data
• Planning ensures that customers can be targeted.	• Inputs are random.
• Questions can be tailored to get precision in data collection.	• Inputs may not have anything to do with the questions that need to be answered.
• Questionnaires can be designed to avoid bias.	• Self-selection bias present; data are collected only from those who interact.
• Quantitative information can be collected.	• Most inputs are qualitative.
• Active data collection asks hypothetical questions.	• Data directly available about context of usage of product or service.
• Hard to capture *emotional* quality of data.	• Interactions capture passions and emotions of the moment.

FIGURE 10.4
Provides a comparison of active versus passive VOC data.

producing, what the customer requirements are, what the customers actually want, and what would excite the customer. Active listening and accurate interpretation of the VOC is essential. Remember, it is important to listen for what they need, not to what they say.

The task of the data collection team is to extract true needs from the customers' statements (see Table 10.2).

It was not long ago that cars did not have cup holders. Then one day, they were introduced by Japanese automakers. This delighted customers at the time. Anyone that had spilled a Big Gulp or a Starbucks onto his or her new carpeting appreciated this innovation. Today, they are an expectation. "What—no cup holder?" can have potential car buyers leaving the lots in droves. What is more, cup holders have become an engineering art form. How do you get all the different-size cups to fit? The innovation is endless.

TABLE 10.2

Customer-Related Data

What They Say	What They Really Need
I want my name to be recorded on a frequent customer database.	As a frequent customer, I need to be recognized.
I want on-time delivery.	I want your delivery to be here when I need it.
I need 99% accuracy in your invoices.	I need the invoice to accurately reflect what I owe.
Your shipments always arrive broken.	I need the goods you send to me to arrive undamaged.

Threshold quality attributes and *excitement* quality attributes are nonlinear—small changes in performance create large changes in satisfaction. *Performance* quality attributes are linear—small changes in performance create proportional changes in satisfaction. Even high performance on threshold quality attributes produces only small amounts of satisfaction; conversely, moderate performance on excitement quality attributes produces large increases in satisfaction. Meeting excitement quality attributes produces the greatest returns per unit of effort. Not meeting excitement quality attributes has the greatest risk. Threshold quality attributes are the *ho-hum* needs; risks and returns are both smaller. The Kano model represents a single snapshot in time—needs evolve from threshold to performance to excitement over time.

EXAMPLES

Example with a car:
- Threshold—brakes that will stop the car in 20 feet when driving at 30 miles per hour
- Performance—fuel economy
- Excitement—heated seats

Example with a hotel experience:
- Threshold—temperature of the water in the shower. Adjustable from 60°F to 110°F
- Performance—timeliness of having clean bed sheets each night
- Excitement—a hot cookie delivered to your room free of charge

SOFTWARE

- http://pragmaticmarketing.com/resources/Prioritizing-Software-Requirements-with-Kano-Analysis
- http://agilesoftwaredevelopment.com/2006/12/kano-model-of-customer-satisfaction

REFERENCE

Kano, N., Seraku, N., Takahashi, F., and Tsuji, S. Attractive quality and must-be quality. *Journal of the Japanese Society for Quality Control* (in Japanese), vol. 14, no. 2, pp. 39–48, 1984.

SUGGESTED ADDITIONAL READING

Cadotte, E.R., Turgeon, N. Dissatisfiers and satisfiers: Suggestions from consumer complaints and compliments (pdf). *Journal of Consumer Satisfaction, Dissatisfaction and Complaining Behavior* vol. 1, pp. 74–79, 1988.

11

Nominal Group Technique

H. James Harrington

CONTENTS

Everyone has his/her own idea of what is most important. The trick is to come to a common agreement on priorities.

H. James Harrington

DEFINITION

Nominal group technique (NGT) is a technique for prioritizing a list of problems, ideas, or issues that gives everyone in the group or team equal voice in the priority setting process.

USER

NGT is a powerful and time-tested group ideation and problem-solving technique involving the so-called triple crown of problem identification, creative solution generation, and decision making. It can easily and consistently be used in groups of many types and sizes—groups or teams who want to make their decision quickly by voting—but who want at the same time everyone's input and opinions taken into account.

OFTEN USED IN THE FOLLOWING PHASES OF THE INNOVATIVE PROCESS

The following are the seven phases of the innovative cycle. An X after the phase name indicates that the tool/methodology is used during that specific phase:

- Creation phase X
- Value proposition phase
- Resourcing phase
- Documentation phase
- Production phase X
- Sales/delivery phase X
- Performance analysis phase

TOOL ACTIVITY BY PHASE

- Creation phase—As a result of a typical brainstorming session, a long list of options/problems is created. After discussing the list, the brainstorming team will need to prioritize the ideas that they will focus their initial effort on. This is where the NGT provides an effective approach to getting common agreement from the team.

- Production phase—Frequently during the production phase, there are a number of alternatives that need to be considered, for example, what would be produced, what parts will be farmed out, which suppliers should be used, where should product be stored, etc. The NGT is frequently used in cases where there are differences of opinion about what option is the best for the organization.
- Sales/delivery phase—During this phase, an innovative team will come up with many different options relating to the marketing sales and delivery approaches. The NGT is frequently used to help gain common agreement on which of the alternatives is the one that will be pursued.

HOW THE TOOL IS USED

NGT is a technique for prioritizing a list of problems, ideas, or issues that gives everyone in the group or team equal voice in the priority-setting process. Each team member on their own determines their top priority options. (For example, if they were to select their top three options, they would rate them their first, second, and third choices.) The individual ratings are recorded behind the item and are weighted based on the individual's priority. The sum of these weightings for the individual items is used to determine the priority of the item.

NGT is a special-purpose technique, useful for situations where individual judgments must be tapped and combined to arrive at decisions. It is a process that is best accomplished by all the people involved in the situation, and is a problem-solving or idea-generating strategy not typically used in routine meetings. Andre L. Delbecq and Andrew H. Van de Ven developed NGT in 1968. It was derived from social-psychological studies of decision conferences, management science studies of aggregating group judgments, and social work studies of problems surrounding citizen participation in program planning. Since that time, NGT has gained extensive recognition and has been widely applied.

NGT takes its name from the fact that it is a carefully designed, structured, group process. It involves carefully selected participants in some activities as independent individuals, rather than in the usual interactive mode of conventional groups. It is a well-developed and tested method, which was fully presented in the work of Delbecq et al. (1975).

This tool is best used when time is limited and when group ownership of the decision is necessary. It is also useful when multiple creative ideas are not necessarily critical to the solution of a problem, or when the focus is on the solution of a problem.

The NGT is a seven-step process, which works best in a group of 7 to 10. The participants are physically present, and a leader or facilitator controls the session.

Steps for Using NGT

Step One

Following an opening introduction in which the purposes of the session are outlined, participants are presented with a carefully worded task statement and NGT worksheet (see Figure 11.1) (Harrington and Lomax, 2000).

Step Two

The group members are then instructed to write, on the worksheet provided, their own ideas for the resolution to the problem statement. This second step is called *silent generation* and typically takes about 10 minutes. No discussion should be allowed during this time.

Nominal group technique worksheet—Ideas	
Problem	**Date:** _____
Individual ideas	

FIGURE 11.1
Example of an NGT idea worksheet.

Step Three

Next comes the *round-robin phase* during which the facilitator, or leader, calls on participants one by one to state one of the responses he or she has written. Participants may pass at any time and join in on any subsequent round. A participant may propose only one item at a time, and either the facilitator, leader, or assistant records each item as it is presented. The only discussion allowed is between the recorder and the participant who proposes the item, and it is limited to seeking a concise rephrasing for ease of recording. Participants are encouraged to add items to their personal list should new ones occur to them during the round robin. This continues until all ideas are recorded.

Step Four

The fourth step is *clarification*. Once all items have been recorded, the facilitator or leader goes over them, one at a time, to ascertain that all participants understand the actions that have been recorded. Duplicate ideas should be eliminated. New ideas can be generated at this time by combining ideas that were generated individually. Any participant may offer clarification, or may suggest combination, modification, deletion, etc., of items. However, no evaluation is permitted. The final list of ideas should be assigned letters, A through Z, where Z depends on how many ideas were generated.

Step Five

In step five, *ranking*, each participant selects the single best-preferred item and gives it the highest rank. The highest rank equals the number of items that the group has decided to rank, for example, 5, or 10, or 20, etc. The remaining ideas are then ranked, with the least preferred item getting the rank of 1. Use the form shown in Figure 11.2.

Step Six

In step six, the leader asks the participants to share their preferred ideas. As each one contributes an idea, the recorder makes a tick mark next to the idea on the flipchart. The idea with the highest ranking is the one that is most favored by the group. This result will be very close to group consensus.

Nominal group technique worksheet—Ranking		
Problem statement:	Date: _____	
Idea		Ranking
A	_____	
B	_____	
C	_____	
D	_____	
E	_____	
F	_____	
G	_____	

FIGURE 11.2
Example of an NGT ranking worksheet.

Step Seven

The final step is for the leader to verify the consensus of the group on the single idea preferred by the entire group and develop action plans on the basis of the overall consensus.

The NGT stimulates generation of ideas by keeping everyone involved in the problem-solving process; it prohibits strong personalities from dominating the group; it allows ideas to be weighted for their relative worth, and it encourages a shared commitment to solutions and implementation.

EXAMPLES

A team is presented with the following task statement: "Our administrative group cannot keep pace with the required documentation generation." Each team member would document his or her response on the ideas worksheet. Team member Bob shows the information in Figure 11.3 on his worksheet.

In phase three, each participant has given their own idea to the team. The facilitator has recorded all ideas on a flipchart (see Figure 11.4).

After our team idea list has been clarified and narrowed, it is time for Bob and the rest of the team to rank the remaining ideas. Bob's ranking form is presented in Figure 11.5.

Nominal group technique worksheet—Ideas

Problem statement: **Date:** January 5, 1998
Our administrative group cannot keep pace with the required
documentation generation.

Individual ideas:
Hire more people
Distribute the work more evenly
Take in fewer new tasks

FIGURE 11.3
Bob's ideas worksheet.

Hire more people
Distribute the work more evenly
Take in fewer new tasks
Change from word processors to computers
Hire part-time workers
Use management to fill-in
Take fewer breaks
Take more breaks
Buy new word processors
Provide better training
Provide any training
Provide comfortable chairs
Designate work-flow supervisors
Use first-in, first-out system
Outsource some of the work

FIGURE 11.4
Team's idea list.

A new team list of *preferred* ideas is generated. Very often, this will consist of each team member's *first* and *second* choices (see Figure 11.6).

Once again, now is the time for Bob's team to reach consensus on which of the solutions to go forward with. In this case, the team selected to change from word processors to computers as a solution to their issue.

Nominal group technique worksheet—Ranking	
Problem statement: **Date:** January 5, 1998 *Our administrative group cannot keep pace with the required documentation generation.*	
Idea	Ranking
A. Hire more people	7
B. Distribute the work more evenly	8
C. Take in fewer new tasks	9
D. Out-source some of the work	6
E. Change from word processors to computers	1
F. Hire part-time workers	3
G. Provide better training	4
H. Use first-in, first-out system	5
I. Buy new word processors	2

FIGURE 11.5
Bob's ranking worksheet.

Team's preferred ideas
Change from word processors to computers Buy new word processors

FIGURE 11.6
Team's preferred ideas.

SUMMARY

As you can see, NGT uses several other basic tools, such as brainstorming, narrowing techniques, and consensus. This tool is very simple to use, yet it is dynamic in its results.

SOFTWARE

No software required.

REFERENCES

Delbecq, A.L., Van de Ven, A.H., and Gustafson, D.H. *Group Techniques for Program Planning: A Guide to Nominal Group and Delphi Processes*. Middleton, WI: Green-Briar-Press, 1975.

Harrington, H.J. and Lomax, K.C. *Performance Improvement Methods*. New York: McGraw-Hill, 2000.

SUGGESTED ADDITIONAL READING

Levi, D.J. *Group Dynamics for Teams*. Thousand Oaks, CA: SAGE Publishing, 2010.

12

Plan–Do–Check–Act (Shewhart Cycle)

H. James Harrington

CONTENTS

Problems breed problems, and the lack of a disciplined method of openly attacking them breeds more problems.

Philip Crosby
Quality Is Free, 1980

DEFINITION

Plan–Do–Check–Act (PDCA) is a structured approach for improving services, products, or processes. It is also sometimes referred to as plan–do–check–adjust. Another version of this PDCA cycle is OPDCA. The added O stands for observation or as some versions say "Grasp the current condition."

USER

This tool can be used by individuals or groups but is best used with a group of four to eight people. Cross-functional teams usually yield the best results from this activity.

OFTEN USED IN THE FOLLOWING PHASES OF THE INNOVATIVE PROCESS

The following are the seven phases of the innovative cycle. An X after the phase name indicates that the tool/methodology is used during that specific phase.

- Creation phase X
- Value proposition phase
- Resourcing phase
- Documentation phase
- Production phase X
- Sales/delivery phase X
- Performance analysis phase

TOOL ACTIVITY BY PHASE

- Creation, production, and sales/delivery phases—The PDCA methodology is extensively used in each of these phases to address problems or opportunities that arise to develop a corrective action plan to eliminate the problem or to take advantage of an opportunity.

HOW THE TOOL IS USED

The PDCA cycle is a very simple approach to project management that can be used effectively on noncomplex programs and for implementing corrective action. Often, it is incorrectly called the *Deming cycle*. It was actually designed by Walter A. Shewhart and first published in his book, *Statistical Method from the Viewpoint of Quality Control* (Shewhart, 1939).

Although Deming always referred to this activity as the *Shewhart cycle*, the Japanese called it the Deming cycle. This was because it was Deming who introduced this process to Japan in 1950. Deming stated:

> The Shewhart cycle was on the blackboard for top management for every conference beginning in 1950 in Japan. I taught it to engineers— hundreds of them—that first hot summer. More the next summer, six months later, and more six months from that. And the year after that, again and again.

We can safely say, while Shewhart invented the approach, Deming became its salesman.

Most people today associate the PDCA activity with the four key steps shown in Figure 12.1.

Steps to Complete the Cycle

Actually, there are five steps required to complete the cycle. Let us take a look at all five. The first part of each step comes from Shewhart's book (Shewhart, 1939).

Step 1. Plan

Objective: To transform objectives of an activity into a clear and logical structure.

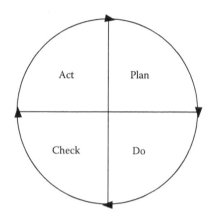

FIGURE 12.1
Shewhart cycle.

> What could be the most important accomplishments of this team? What changes might be desirable? What data are available? Are new observations needed? If yes, plan a change or test. Decide how to use the observations.

Begin by studying an organization's current situation. Before any improvement plans are made, ensure that the current best-known methods are documented and standardized. It is imperative to start from a stable base so the effectiveness of actions can be evaluated later.

Next, gather data to identify and define the problem(s) and to help formulate a plan. Only then can planning be initiated for the desired accomplishments over a given period of time and for how the effect of the planned actions will be systematically measured. The plan should include specific actions, changes, or tests that are the outgrowth of a systematic study of the probable causes of the problem(s) or effect(s) in question using statistical methods and problem-solving tools.

Step 2. Do

Objectives: Collect data, indicators, and information required to take action to complete the assignment.

> Carry out the change or test decided upon, preferably on a small scale. (Also, search for all available data, which may assist in answering the questions in step 1.)

Implement the plan. If possible, try it out on a small scale first. Insist that all relevant changes are recorded during implementation, and any changes from the planned measures are documented. Ensure that data are collected systematically and in a way that facilitates evaluation (e.g., use checksheets).

Step 3. Check

Objectives: Evaluate the results to determine if the objectives of the activity were accomplished.

Observe the effects of the change or test.

Evaluate the data collected during implementation to see if the desired objectives were met. Check the results to see if there is a good fit between the original goals and what was actually achieved.

Step 4. Act

Objectives: If the desired results were obtained, ensure that the appropriate people are trained and the new processes are adequately documented. If the desired results were not accomplished, implement restarts to the PDCA system to accomplish the desired results (start step 5).

Study the results. What did we learn? (Also, what can we predict?)

Depending on the results of the previous evaluation, take further actions. If successful, adopt the changes. That is, institutionalize the change taken by documenting the new procedures, communicating them to all personnel in the process, and training people to the new standards. The new methods, procedures, and specifications then can be replicated in all areas with similar processes. If unsuccessful, abandon the changes.

Step 5. Repeat Steps 1 through 4

With the knowledge accumulated, steps 1 through 4 can be repeated to bring about an even greater improvement.

The following shows the 10 detailed tasks that together expand on the activities that take place during the PDCA methodology.

Plan

1. Identification of targets (objectives and goals)
2. Identification of methods/procedures to achieve these targets
3. Identify control items and methods

Do

4. Communicate and train associates
5. Implement the plan (2 and 3)

Check

6. Check progress to plan (1, 2, and 3)
   - Against targets and goals
   - Within the strategy
7. Identify any problems

Act

8. Resolve/eliminate problems
9. Correct/modify plan (2 and 3)
10. Standardize the improvement

SUMMARY

Organizational improvement will last only if it is a continuous process. Standards and processes, in both manufacturing and business, must be constantly reviewed, improved or reengineered, measured, and monitored. Continuously passing them through Shewhart's PDCA cycle is an effective way of driving the continuous improvement process. Most of the popular problem-solving techniques are based on Shewhart's PDCA methodology. For example, the Six Sigma approach define–measure–analyze–approve–control methodology is just a derivation of Shewhart's methodology.

EXAMPLE

The Shewhart cycle has been used in many ways. The environmental group that designed ISO 14001 built it around the Shewhart cycle (see Figure 12.2).

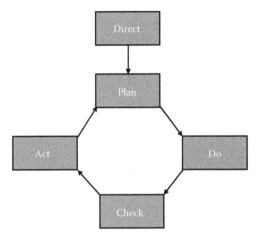

FIGURE 12.2
ISO 120001 pattern.

While the original Shewhart cycle begins with the *planning* phase, the ISO 14001 pattern begins by setting directions to the organizations—then progresses through the rest of the PCDA cycle. The following is a list of all the ISO 14001 clauses that apply to each of the five categories.

Direct
    4.1.0 General
    4.2.0 Environmental policy
Plan
    4.3.0 Planning
    4.3.1 Environmental aspects
    4.3.2 Legal and other requirements
    4.3.3 Objectives and targets
    4.3.4 Environmental management programs
Do
    4.4.0 Implementation and operation
    4.4.1 Structure and responsibility
    4.4.2 Training, awareness, and competence
    4.4.3 Communication
    4.4.4 Environmental management system documentation
    4.4.5 Document control
    4.4.6 Operational control
    4.4.7 Emergency preparedness and response

Check

4.5.0 Checking and corrective action

4.5.1 Monitoring and measurement

4.5.2 Nonconformance and corrective and preventive action

4.5.3 Records

4.5.4 Environmental management system audit

Act

4.6.0 Management review

SOFTWARE

- Plan–Do–Check–Act: Software 3.2, description by sqaki.com
- SAT/ACT/PSAT Platinum 2012—$49.95
- SAT/PSAT/ACT Bootcamp—$19.95

REFERENCE

Shewhart, W.A. *Statistical Method from the Viewpoint of Quality Control*. New York: Dover, 1939. ISBN 0-486-65232-7.

SUGGESTED ADDITIONAL READING

Deming, W.E. *Out of the Crisis*. Cambridge, MA: MIT Center for Advanced Engineering Study, 1986. ISBN 0-911379-01-0.

Harrington, H.J. and Lomax, K.C. *Performance Improvement Methods*. New York: McGraw-Hill, 2000.

Levi, D.J. *Group Dynamics for Teams*. Thousand Oaks, CA: SAGE Publishing, 2010.

Shewhart, W.A. *Economic Control of Quality of Manufactured Product/50th Anniversary Commemorative Issue*. Milwaukee, WI: American Society for Quality, 1980. ISBN 0-87389-076-0.

13

Reengineering/Redesign

H. James Harrington

CONTENTS

To have an innovative organization, you have to be constantly applying innovation to the organization's processes.

H. James Harrington

DEFINITION

- Process—a series of interrelated activities or tasks that take an input and produce an output.
- Process redesign—a methodology used to streamline a current process with the objective of reducing cost and cycle time by 30% to 60%, while improving output quality from 20% to 200%.
- Process reengineering—a methodology used to radically change the way a process is designed by developing an aggressive vision of how it should perform and using a group of enablers to prepare a new process design that is not hampered by the present process paradigms. Use when a 60% to 80% reduction in cost or cycle time is required. Process reengineering is sometimes referred to as *new process design* and/or *process innovation*.

USER

This tool can be used by a group of 4 to 12 people. Cross-functional teams usually yield the best results from this activity.

OFTEN USED IN THE FOLLOWING PHASES OF THE INNOVATIVE PROCESS

The following are the seven phases of the innovative cycle. An X after the phase name indicates that the tool/methodology is used during that specific phase:

- Creation phase X
- Value proposition phase X
- Resourcing phase
- Documentation phase X
- Production phase X
- Sales/delivery phase
- Performance analysis phase

TOOL ACTIVITY BY PHASE

As indicated above, reengineering/redesign methodologies are primarily used during the creation, value proposition, documentation, and production phases:

- During the creation phase, the methodology is used to create the future-state definition of the process being studied.
- During the documentation phase, the future-state definition is documented in operating procedures and flowcharts.
- During the production phase, the future-state solution is implemented and adjustments are made to optimize overall performance.

HOW TO USE THE TOOL

Process improvement methodologies have evolved rapidly during the last 90 years. This evolution reflects the changing attitude from people being the problem to making the processes capable of producing excellent error-free results, while minimizing costs and cycle time. Figure 13.1 shows some of the key highlights of the process improvement evolution.

It is very important to understand that both process redesign and process reengineering are very different than a continuous improvement model. Continuous improvement focuses on refining and improving the present state (see Figure 13.2), while both process redesign and process

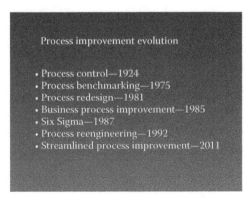

Process improvement evolution

- Process control—1924
- Process benchmarking—1975
- Process redesign—1981
- Business process improvement—1985
- Six Sigma—1987
- Process reengineering—1992
- Streamlined process improvement—2011

FIGURE 13.1
Evolution of process improvement methodologies.

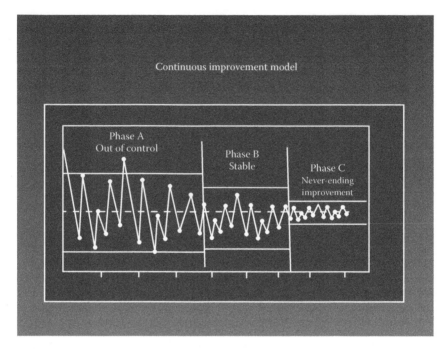

FIGURE 13.2
Impact of continuous improvement.

reengineering focus on developing a new performance standard and the new future-state solution (see Figure 13.3).

Process redesign and process reengineering are two very different methodologies. Although they both focus on improving process performance, they use very different methodologies. As a result, in this chapter, we will separate them into two parts.

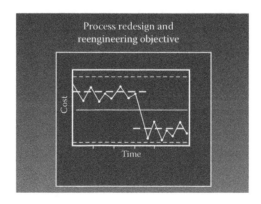

FIGURE 13.3
Process output objectives of process redesign and process reengineering.

Process Redesign

Redesign requires out-of-the-box thinking to change the process to the point where it sets new benchmark standards.

H. James Harrington
Process Management Excellence, Paton Press, 2006

The Process redesign activities started in IBM in the late 1980s when the executive management team set an objective to not let Japan influence IBM's business reputation like Japan had influenced GM and Ford. The team that was assigned to lead this initiative realized early in the process that the best way to keep IBM as the leader in the computer industry was to focus on having innovative processes in all parts of the organization. All major processes had process owners assigned to them with the sole responsibility of optimizing the performance of these processes using creative and innovative approaches. By the early part of the 1980s, IBM rolled out a program called *business process improvement.* Today, this is commonly called *process redesign*, as it was primarily focused on using information technology along with streamlining all major processes to make them more efficient effective and adaptable (Harrington, 1991).

The process redesign methodology takes an existing process and removes waste while reducing cycle time and improving the product's effectiveness. After the process is simplified, automated and information technology are applied, maximizing the process' ability to improve efficiency and effectiveness, and the adaptability measurements. Process redesign is sometimes called focus improvement because it concentrates effort on the present process. It results in improvements in efficiency that range between 30% and 1000%, and reduces cost and cycle time by 30% to 60%. Process redesign is the breakthrough methodology most frequently used because the risks are low and the costs are less. This is the right answer for approximately 70% of business processes. The redesign process consists of five phases (see Figure 13.4). This process is called PASIC, which stands for:

P—planning, organizing for improvement
A—analyzing, understanding the present process
S—streamlining the process
I—implementing, measuring, and controlling the future state process
C—continuously improving, ongoing refinement of the new process

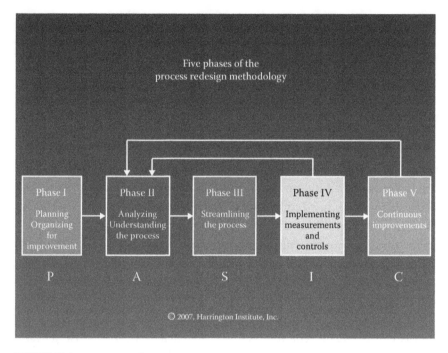

FIGURE 13.4
Five phases of the process redesign methodology.

People who are working on refining and developing new processes quickly realize that the major problems occur at the handoff between people, activities, departments, and functions. As a result, a great deal of the process redesign activities are focused on process alignment as pictured in Figure 13.5 (Harrington, 2006).

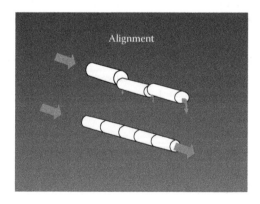

FIGURE 13.5
Process alignment.

- Phase I. Planning for improvement—Consists of eight different activities (see Figure 13.6). This phase is directed at selecting the right processes to improve, organizing and training the team to create the future-state solution, establishing realistic goals and objectives for the project, and establishing an approved project plan.
- Phase II. Analyzing the present process—This phase consists of eight activities (see Figure 13.7). During this phase, a number of process flow diagrams are constructed that depict activity flow, information flow, product flow, etc. These are validated by a process walk-through where hard factual data is collected. Often, simulation models are prepared so that accurate projections over time can be made as conditions change.

Typical diagrams used during phase II are:

- Block diagram
- ANSI standard
- Geographic
- Functional
- Data flow

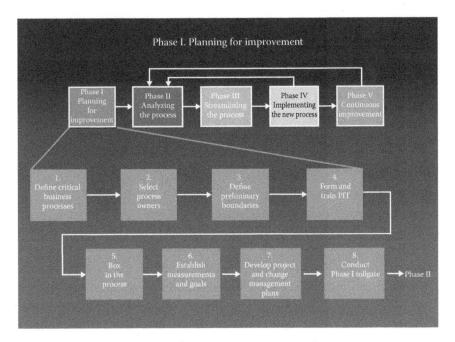

FIGURE 13.6
Phase I of PASIC methodology.

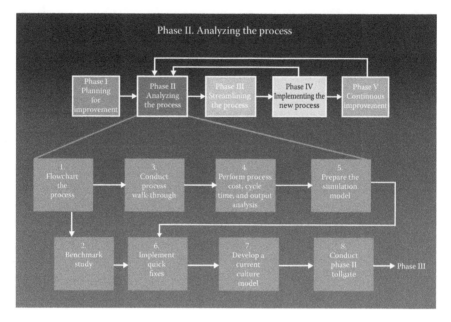

FIGURE 13.7
Phase II of PASIC methodology.

- Communication flow
- Knowledge flow
- Value flow
- Value stream
- Administrative business process principle diagrams
- Hierarchical overview diagram
- Global overview of process and division diagrams
- Global process diagram
- Detailed process diagrams
- Form management diagrams
- Form circulation diagrams
- Accounting system diagrams
- Etc.

Figures 13.8 and 13.9 are typical examples of diagrams that are constructed during this phase. Figure 13.10 is a typical example of the diagram after phase III is completed and the processes have been streamlined.

Figure 13.10 is a future-state map of the process using the value proposition approach to diagramming.

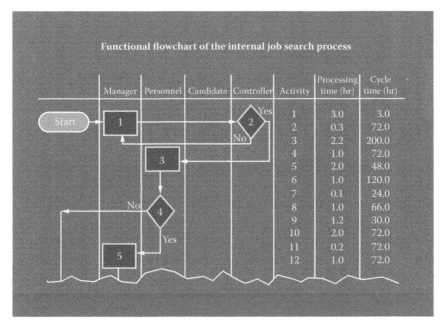

FIGURE 13.8
Typical first part of a functional flowchart, sometimes called a swim lane flowchart.

Example of a current state value stream map

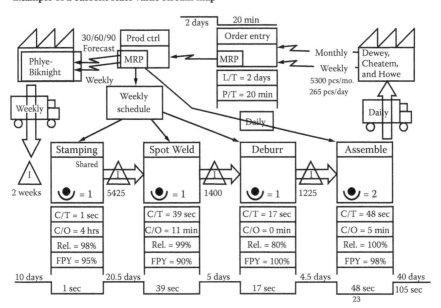

FIGURE 13.9
Current-state map of the process using the value proposition approach to diagramming.

Example of a future state value stream map

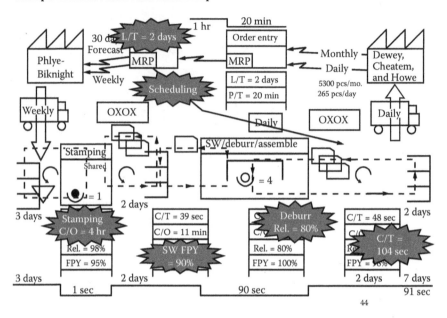

FIGURE 13.10
Future-state map that reflects the process after it has been streamlined.

- Phase III. Streamlining the process—During this phase, the process improvement team (PIT) makes effective use of its creative and innovative knowledge. This phase consists of six activities (see Figure 13.11).

  The heart of all of the PASIC methodology rests in activity 1 of phase III: *Applying streamlining approaches.* Activity 1 includes 12 tasks (see Figure 13.12). These tasks start with focusing on eliminating bureaucracy (task 1) and end with simplifying language (task 12).

  In task 2 (value-added analysis), the PIT creates a rainbow flowchart that highlights in color the operations that are bureaucracy (blue), business value-add (yellow), no value-add (red), and value-add (green) (see Figure 13.13). (Since this book is not printed in color, you can't visualize the various colors depicted in the Rainbow flowchart. However, we will explain how a Rainbow Flowchart would be created in your work environment.) You should color in the individual different types of activities in the flowchart in order to get in the immediate picture of where the potential problems are occurring.

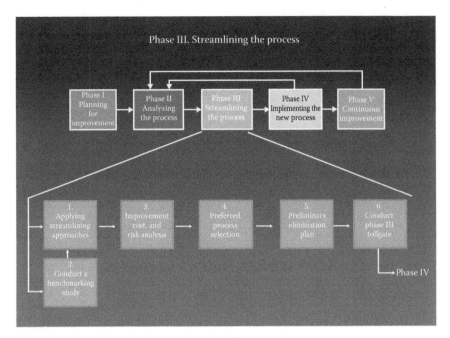

FIGURE 13.11
Phase III of PASIC methodology.

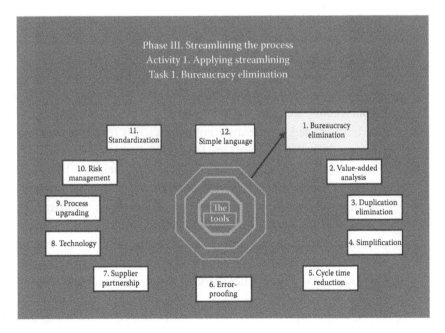

FIGURE 13.12
Twelve tasks included in applying the streamlining activities to the process.

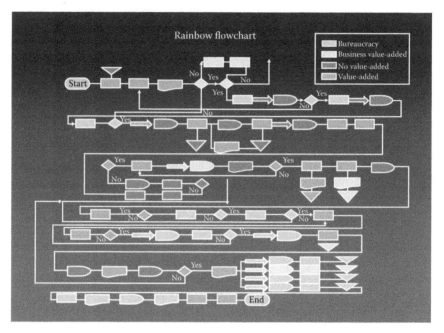

FIGURE 13.13
Typical rainbow flowchart.

To do this, we colored in the bureaucracy with one color (blue). We colored in the no-value-added activities a different color (red), the business-value-added activities should be colored in with a different color (yellow). The real-value-added activities were left white. This creates what we call a Rainbow flowchart. We recommend that you stand back about 20 feet from the flowchart so you're not reading the words, but focusing on the different colors. This technique quickly highlights where the problems are in the process.

Often, depending on the skills of the PIT, activity 1 will be repeated one or two more times using different control parameters. For example, the first time, there are no restrictions on the team. The next time, the new design must be installed and operating within 90 days. The third time, the cost of installing the new process cannot be more than 25% of the cost of the first proposed new process. This frequently leads to a fourth new process design that includes the best features of the other three designs. All too often, the team members spend all of their time justifying the process they first designed, and do not search for other solutions that may give better results. This approach makes the PIT look at other solutions with an open mind. All too often, we have an idea and hold onto it without searching for a better solution. The results from this approach provide the executive team with a view of different ways to improve the process, and allow them to have a selection. Typically, the selection is very obvious once we have looked at the alternatives. Figures 13.14 and 13.15 are typical examples.

- Phase IV. Implementation, measures, and controls—This phase consists of five activities (Harrington, 2012):
  1. Final implementation plan
  2. New process implementation
  3. In-process measurements
  4. Feedback system installed
  5. Poor-quality costs system installed

Be sure that the change management part of your project plan has been fully implemented, and the affected personnel are ready for the radical change in the process and the structure. If not, do not start to install the new process until the people affected have been adequately involved so that they will accept any radical changes as necessary.

Performance estimate				
	Original process	Case 1	Case 2	Case 3
Effectiveness (quality)	0.2	0.02	0.01	0.009
Efficiency (productivity)	12.9 hr/cycle	7.5 hr/cycle	6.3 hr/cycle	5.3 hr/cycle
Adaptability measurement	25%	Not measured	80%	65%
Cycle time	305 hr	105 hr	105 hr	85 hr
Cost per cycle	$605	Not measured	$410	$380

© 2007, Harrington Institute, Inc.

FIGURE 13.14

Comparison of projected performance to each other and the present process for three alternative cases.

Implementation estimate

	Case 1	Case 2	Case 3
Cost	$1,300,000	$20,000	$280,000
Implementation cycle time	24 months	6 months	15 months
Probability of success	50%	95%	85%
Major problem	Need more data	Training time	New org. structure

FIGURE 13.15
Comparison of costs to each other for three alternative cases.

- Phase V. Continuous improvement—During this phase, the process parts are turned over to the individual managers responsible for that part of the process. These managers are assigned the responsibility to improve their part of the process an average of 5% to 10% a year. The process owner still is held responsible for the total process and to ensure the suboptimization does not occur. The team is assigned to improve the process so that it will systematically progress up the six levels of the process maturity grid (see Figure 13.16).

Process Reengineering

Reengineering is radical—revolutionary rather than evolutionary.

H. James Harrington

Do not confuse process reengineering with reverse engineering:

- Process reengineering is a methodology used to radically change the way a process is designed by developing an aggressive vision of how it should perform and using a group of enablers to prepare a new process design that is not hampered by the present process

Six-level process maturity grid

Level	Status	Description
6	Unknown	Process status has not been determined.
5	Understood	Process design is understood and operates according to prescribed documentation.
4	Effective	Process is systematically measured, streamlining has started and end-customer expectations are met.
3	Efficient	Process is streamlined and more efficient.
2	Error-free	Process is highly effective (i.e., error-free) and efficient.
1	World-class	Process is world-class and continues to improve.

FIGURE 13.16
Six levels of the process maturity grid.

paradigms. Use when a 60% to 80% reduction in cost or cycle time is required. Process reengineering is sometimes referred to as new process design or process innovation.

- Reverse engineering is a periodic purchasing of a product from competitors to test and disassemble along with a correlation sample of the organization's product. Typical points to be analyzed include order cycle time, packaging protection, installation instructions, product characteristics, initial performance, reliability, safety factors, environmental performance, suppliers, ease to repair, assembly methods, workmanship, and cost to produce.

Process reengineering is the most radical of the process breakthrough methodologies. It is sometimes called *process innovation* because its success relies heavily on the PIT's innovation and creative abilities. Other organizations call it *big picture analysis* or *new process design*. We like the term *new process design* best because the approach used is the same as if the organization were designing the process for the first time. This approach takes a fresh look at the objectives of the process and completely ignores the present process and organizational structure. It starts with a blank sheet of paper as you would if you were engineering the process for the first time.

Process reengineering, when applied correctly, reduces cost and cycle time between 60% and 90% and improves quality from 20% to 100%. It is a very useful tool when the current-state process is so out of date that it is not worth salvaging or even influencing the best-value future-state solution. Process reengineering is the correct answer for 5–15% of the major processes within an organization. If you find it advantageous to use process reengineering in more than 20% of your major processes, the organization should be very concerned, as it may be indicative of a major problem with the management of the organization. This management problem should be addressed first, before a great deal of effort is devoted to improving processes that will not be maintained.

The process reengineering approach allows the PIT to develop a process that is as close to ideal as possible. The PIT steps back and looks at the process with a fresh set of eyes, asking itself how it would design this process if it had no restrictions. This approach takes advantage of the available process enablers, including the latest mechanization, automation, and information technology techniques, and improves upon them. Often, this process stimulates the PIT to come up with a radical new process design that is truly a major breakthrough.

The process reengineering approach provides the biggest overall improvement, but it is the most costly and time-consuming approach. It also has the highest degree of risk associated with it. During the process reengineering cycle, the PIT will challenge all of the organization's sacred concepts and its paradigms related to the process being evaluated. The PIT is empowered to make "hamburger out of any sacred cow" that stands in their way of creating the ideal process. Often, the process reengineering approach includes organizational restructuring and can be very disruptive to the organization. Most organizations can only effectively implement one change of this magnitude at a time.

> Reengineering as a term carries negative connotations (terminations, downsizing, layoffs). We don't use it. Instead we focus on customer satisfaction and value-added processes.
>
> **Controller, $10 billion paper goods firm**

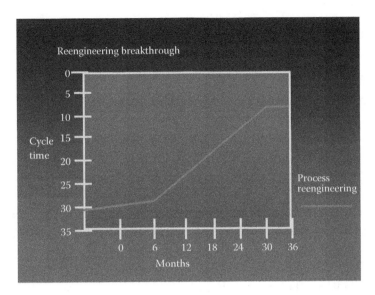

FIGURE 13.17
Typical process reengineering improvement cycle.

Figure 13.17 represents the typical impact that a process reengineering project has on cycle time. You will note that at the starting point, it took 30 days to go through the cycle. At the end of 36 months, the cycle time had been reduced to eight days.

The process reengineering approach to developing a best-value future-state solution consists of five tasks:

Task 1. Big picture analysis
Task 2. Theory of ones
Task 3. Process simulation
Task 4. Process modeling
Task 5. Install the new process

Task 1: Big Picture Analysis

With this task, the PIT is not constrained in its vision. The results of the process reengineering activities must be in line with the corporate mission and strategy. They should also reinforce the organization's core capabilities and competencies. Before the PIT starts to design the new process, it needs to understand where the organization is going, how the process being evaluated supports the future business needs, and what changes would provide the organization with the most important competitive advantage.

Once this is understood, the PIT can develop a vision statement of what the best process would look like and how it would function. In developing the vision statement, the PIT needs to think outside of the normal routine (thinking outside the box) and challenge all assumptions and all constraints, question the obvious, identify the technologies and organizational structures that are limiting the process, and define how these factors can be used to create processes that are better than today's best. The vision statement defines only what must be done, not what is being done. Usually, the vision statement is between 10 and 30 pages and, in reality, is more like a new process specification. The vision statement includes definitions of what all of the stakeholders would like the process to look like:

- High-level process descriptions
- List of all the potential people enablers
- List of all the potential technology enablers
- List of all the potential process enablers
- List of all the potential organizational change enablers
- List of all the potential organizational structure enablers
- Projected performance specifications
- Assumptions
- List of critical success factors

Task 2: Theory of Ones

Once the vision statement has been finalized, the PIT should define what must be done within the process from input to delivery to the customer. It needs to question why the process cannot be done in one activity by one person in one place, or, better still, at one time with no human intervention. The PIT should be a miser in adding activities and resources to the process. To accomplish this, the PIT will use an approach called the *theory of ones* (see Figure 13.18).

To use the theory of ones, the PIT sets the minimum quantity of units that the organization is trying to optimize. For example, if the PIT is interested in optimizing cycle time and the previous cycle time was five days, it might ask the questions: "What if I had to do it in 1 second? What enablers would

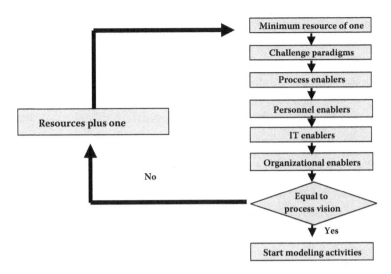

FIGURE 13.18
Theory of ones.

have to be used, and what paradigms would have to be discarded to accomplish this?" Basically, four sets of enablers are addressed:

- Process enablers
- Personnel enablers
- Information technology enablers
- Organizational enablers

After the PIT has looked at each of the enablers to define how they could be used to create a new process that would accomplish the desired function, the resulting process is then compared to the vision statement. If the PIT gets an acceptable answer, it goes forward. If not, it repeats the cycle with the objective of doing the total process in 1 minute. At some point in time, the process and vision statement will be in harmony. As you can see, reengineering is very much an iterative process. You will note that in the process redesign methodology, we look for ways to remove resources and waste. In the process reengineering methodology, we justify why it is necessary to add resources to the process.

Task 3: Process Simulation

When the new process design (future-state solution) is theoretically in line with the objectives set forth in the vision statement, a simulation model is constructed. The simulation model is then exercised to evaluate how the future-state solution will function. If the simulation model proves to be unstable or produces unsatisfactory results compared to the requirements defined in the vision statement, the PIT reinitiates the theory of ones' activity. Then, the PIT prepares and exercises a new simulation model. This cycle is repeated until an acceptable simulation model is constructed.

Task 4: Process Modeling

Once the simulation model indicates that the future-state solution will meet the vision statement, the theoretical model is physically constructed to prove the concepts. Typically, the future-state solution will be evaluated as follows:

- Conference room modeling (without computerization support) to verify the soundness of the new process design
- Pilot modeling in an individual location or small part of the total organization to prove the details of the concepts one at a time
- Pilot modeling of the entire process in a small part of the total organization

Task 5: Install the New Process

With the successful completion of the pilot study, the future-state solution is ready for general roll-out.

EXAMPLES

Now let us assume that we are applying process redesign and process reengineering to an old-fashioned roller skate. If we left it up to the standard continuous improvement process, the results we get are shown in Figure 13.19. The results we got are certainly much better, and that reflects the evolutionary thought pattern that we find in most of the continuous improvement activities.

Now let us take the process redesign approach to bringing about step function improvement in the present-state roller skate. Figure 13.20 could be what the output would look like.

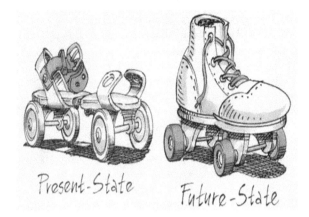

FIGURE 13.19
Output from a present-state roller skate continuous improvement methodology.

FIGURE 13.20
Output from a present-state roller skate process redesign methodology.

Now let us take the process reengineering approach to bringing about radical change to the present-state roller skate. Figure 13.21 could be what the output would look like.

Typical examples of savings:

1. IBM's billing system
   - Fourteen major cross-functional activities located in 255 marketing branches, 25 regional branches, several headquarter departments, and many manufacturing sites

FIGURE 13.21
Roller skate transformed using the process reengineering methodology.

- Of invoices delivered to customers, 98.5% were correct
- Of all invoices, 96% were correct
- Poor-quality cost was 54%

2. Federal-Mogul
   - Reduced development process from 20 weeks to approximately 20 business days, resulting in a 75% reduction in throughput time
3. City of San Jose: Grading Permit
   - Reduced normal permit approval cycle by 76%
   - Express cycle by 90%

SOFTWARE

There are many software involved in the process redesign and reengineering methodologies. Some of the typical ones are for technical enablers (see Figure 13.22).

- Smart Draw: $297–$597, smartdraw.com
- Lucidchart: free, lucidchart.com
- Flowchart Maker: microsoftstore.com

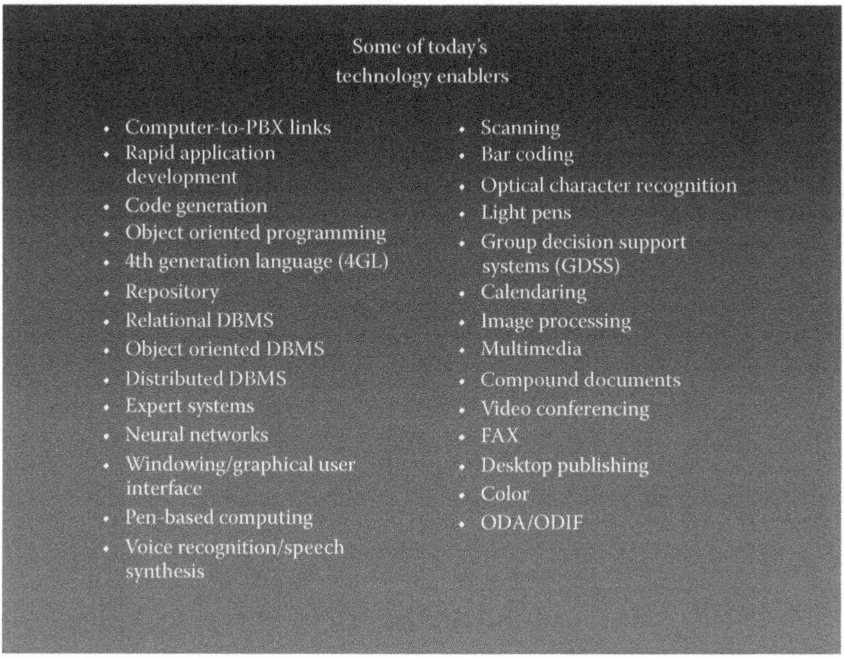

FIGURE 13.22
Technical enablers.

- Microsoft Visio 2013: $300, microsoftstore.com
- MindMap: www.novamind.com/
- QI macros: http://www.qimacros.com

REFERENCES

Harrington, H.J. *Business Process Improvement*. New York: McGraw-Hill, 1991.
Harrington, H.J. *Process Management Excellence*. Chico, CA: Paton Press LLC, 2006.
Harrington, H.J. *Streamlined Process Improvement*. New York: McGraw-Hill, 2012.

SUGGESTED ADDITIONAL READING

Asaka, T. and Ozeki, K., eds. *Handbook of Quality Tools: The Japanese Approach*. Portland, OR: Productivity Press, 1998.
Harrington, H.J. and Harrington, J.S. *High Performance Benchmarking—20 Steps to Success*. New York: McGraw-Hill, 1995.
Harrington, H.J. and Harrington, J.S. *Total Improvement Management*. New York: McGraw-Hill, 1995.
Harrington, H.J. and Lomax, K. *Performance Improvement Methods*. New York: McGraw-Hill, 2000.

14

Reverse Engineering

H. James Harrington

CONTENTS

DEFINITION

This is a process where organizations buy competitive products to better understand how the competitor is packaging, delivering, and selling their product. Once the product is delivered, it is tested, disassembled, and analyzed to determine its performance, how it is assembled, and to estimate its reliability. It is also used to provide the organization with information about the suppliers that the competitors are using.

USER

Usually, a team of reverse engineering specialists is assigned to study how the competitor's product is procured, measure its performance, conduct tests to estimate its reliability, and disassemble the product to understand how it is manufactured and who its suppliers are.

OFTEN USED IN THE FOLLOWING PHASES OF THE INNOVATIVE PROCESS

The following are the seven phases of the innovative cycle. An X after the phase name indicates that the tool/methodology is used during that specific phase.

- Creation phase X
- Value proposition phase X
- Resourcing phase
- Documentation phase
- Production phase
- Sales/delivery phase
- Performance analysis phase

TOOL ACTIVITY BY PHASE

- Creation phase—Reverse engineering activities usually take place as part of the development phase of a new product, but its information is also used in the value proposition phase and the production phase.
- Value proposition phase—The data collected during reverse engineering activities is often used to analyze the competitive advantage a new product will have over its competitors.

HOW TO USE THE TOOL

One of the most effective ways to understand and be able to predict what your competition is going to do is by benchmarking the competitive products. This type of benchmarking is called *reverse engineering*. It is one of the best sources of competitive reliability and design data that is available.

One of the best ways to define opportunities for creating new products is to understand the strengths and weaknesses of your competitors' products. Reverse engineering is one tool that provides you firsthand insight into your competitors' products.

Most organizations are still willing to share information about their business processes. But this is not normally the case when an organization tries to benchmark competitor's hardware, software, customer performance, customer-related services, manufacturing methods, and product design approaches. Even some manufacturing processes and performance data are treated very confidentially by most organizations. Although a lot can be learned through a literature search, contact with appropriate consumer groups, and discussion with subject matter experts, there is nothing more meaningful than firsthand observation, testing, and dissection. For this reason, items are often purchased for competitive product benchmarking.

Reverse Engineering Tasks

A reverse engineering process consists of 11 tasks:

- Task 1. Obtain competitive products
- Task 2. Analyze order and delivery cycle
- Task 3. Analyze packaging and documentation
- Task 4. Characterize the benchmark item(s)
- Task 5. Perform life tests
- Task 6. Perform safety factor analysis
- Task 7. Perform environmental tests
- Task 8. Compare performance results
- Task 9. Perform a product disassembly analysis (reverse engineering)
- Task 10. Compare product design and production methods
- Task 11. Define competitor's competitive advantage

Tasks 1 to 11 are unique to the competitive product benchmarking analysis activities often referred to as reverse engineering (Harrington, 1996). Motorola, for example, used reverse engineering in developing its mobile phones and Bandit pager. When Ford Motor Co. began to design its Taurus model, it disassembled about 50 different mid-size cars from around the world to define each car's best features and assembly methods. There is a competitive evaluation laboratory in one corner of Xerox's Webster plant where, at almost any time, you will see from 20 to 30 competitors' products carefully disassembled, with each of their parts characterized.

There may seem to be something unethical about obtaining competitive products with the sole objective of comparing them to yours, but it is done all the time. In fact, if you are not doing competitive product benchmarking, you are not providing your organization with all the information it needs to make the very best product decisions. As long as the product is available to the general public, it is a candidate for competitive evaluation (Harrington and Harrington, 1996).

Apple Computer Inc. came out with its first portable computer weighing 18 pounds in 1990, only to have Compaq Computer Corp. come out with a notebook computer that weighed only 6 pounds. After disassembling Compaq's notebook computer, Apple engineers were surprised to find out that they could not make an equivalent product. This triggered a major catch-up project that resulted in Apple introducing in 1991 its own notebook computer, weighing between 5.1 and 6.8 pounds depending on the configuration.

The ethical issue is not collecting the data, but how you use it. If you use the data to set performance goals, there is no problem. If you use it to copy the design, you may be infringing on patents and your organization may run into legal problems. There is a fine line between using competitive product benchmarking data to improve your design, and copying a competitor's design. In performing competitive product benchmarking, be careful not to infringe on patents when you are implementing your corrective action.

Note: It is not the intent of this chapter to provide the reader with specific life, stress, or environmental test recommendations. The correct tests for each product must be adjusted to the individual product. This chapter only identifies typical tests that might be performed under specific conditions.

Details of Task 1 through Task 11

- Task 1—Obtain competitive products

  There are two options for obtaining competitive products. Some organizations order products directly from their competitors. (I was always surprised at how many of IBM's new product's first-month's production was delivered to direct competitors.) The other option is to buy the item from a distributor. There are points in favor of both. If you buy directly from your competitor, everything is open and aboveboard. It also allows the buyer to evaluate the competitor's order-processing and delivery activities. The disadvantage is that the competitor can select the sample that they send to you, providing you with biased results. The other option of buying from a distributor ensures that the benchmarking organization receives a random sample of their competitor's product. This is acceptable as long as the distributor also provides the competitor's products to other organizations. Never hire a third party to buy a competitor's products with the objective of keeping the competitor from knowing you have the product. If you buy products from a distributor, you lose the ability to evaluate the competitor's order-processing activities.

- Task 2—Analyze the order and delivery cycle
  As you prepare to place the order, consider all the evaluations you plan on conducting so that a large-enough sample size is ordered. This sample size should already be specifically defined in the benchmarking plan. If the sample is large enough, give consideration to dividing the order up and submitting it at different times. This will help you obtain a more accurate picture of the competitor's product process capabilities. Often, products from different lots or setups perform very differently. When you place the order, keep detailed records related to key performance items. (For example, how many times the phone rings before it is answered, the length of time required to input the order, etc.) Ask for a very short delivery date, one that the reverse engineering team (RET) believes is not possible to meet. This will allow the RET to evaluate how special requests are handled and will also provide the RET with the competitor's normal cycle time. Be sure to record the promised delivery date so that the target and actual order cycle can be calculated. When the product is delivered, check things like
  - How and by whom was the package delivered?
  - Was the package damaged?
  - Was there anything that showed how much the shipping charges were?
- Task 3—Analyze the packaging and documentation
  Be very careful when you unpack the item. Use a video camera to record the total activity. Record the type and weight of all the packing materials, and how the package was organized. Ask yourself how well the item was protected. Evaluate the container to determine how easy it would be for the customer to remove the item from the container without damaging the item. Visually inspect the item to

ensure that it is not damaged in any way. Count all the items that should be in the package to be sure they are there.

Often, the level of protection the packing material provides will also be measured. Units are repackaged using the original packing material and subjected to eight-corner drop tests, incline plane shock tests, and vibration tests. Following each test, the item is unpacked, functionally tested, and visually inspected for damage. Normally, this is an evaluation that is among the last tests done. Often, the item has successfully completed one of the performance tests before this evaluation.

Review the accompanying documentation to determine if it is adequate, if the safety issues are well covered, how the warranty is handled, etc. Analyze the documentation to determine what educational grade level it is prepared for, and if the grade level of the written documentation is in keeping with the potential customer education level. If the item has to be assembled by the customer, follow the assembly instructions exactly to assess how adequate they are and how easy the item is to assemble. Record how many different tools are required to complete the assembly. Ease of assembly is a very important consideration for most consumers.

- Task 4—Characterize the benchmark item(s)
Now is the time to characterize a control sample of your product and the competitor's product. We like to measure one of the competitor's items, then one of the RET's items, to ensure the measurement processes are equivalent. Variable data should be recorded whenever possible, even if your normal practice is to use go–no–go measurement methods. It is well known that differences in distribution can make a big difference in both short- and long-term performance.

For example, when one of the big U.S. auto companies' shipping schedule called for too many gear boxes to be built at their U.S. facility, they turned to a supplier in Japan that provided them with gear boxes manufactured and assembled to the U.S. specifications. When the U.S. auto company compared the field performance of the parts manufactured in Japan to their own, they found that the Japanese product's reliability was much better. As a result, they decided to benchmark the Japanese supplier's product.

To accomplish this, they disassembled a group of the Japanese gear boxes and a control sample of their own, carefully checking the adjustments and measuring each component. Both the Japanese supplier and the U.S.-manufactured parts all met specifications. Further examination of the two sets of data revealed that although the U.S. parts met specification, they varied from one extreme of the specification to the other. In fact, in most cases, it was obvious that parts had been screened, causing a truncated distribution. On the other hand, the Japanese parts were all closely grouped around the center of the specification, using up no more than 50% of the total tolerance.

The lesson they learned is that all parts within a specified tolerance are not equal. Parts that are close to the designed theoretical center point are best, and as they move further away from the center point, they are more susceptible to failure.

To characterize the product, test the product to its acceptance specification. Put the data into the database and compare the initial quality of the competitor's product to the control sample. Any product that does not meet the engineering specification should be dropped from the evaluation at this point.

• Task 5—Perform life tests

A sample of the competitor's product and the control sample should be put to a life/wear test. Exact tests performed differ based on the product. If it is a switch, it could be switched on and off at maximum voltage rating plus 10%, until a failure occurs. A motor could be tested at maximum load, cycling it up to maximum speed and then turning it off, allowing it to cool down before the next cycle starts.

Life tests vary widely from product to product and how the customer will be using the product. Often, stress tests are used to reduce the time to failure. Although this method does not give precise mean time-to-failure data, frequently used stress tests can provide accurate estimates of mean time to failure, and with the use of the control sample, can provide effective comparisons.

When a failure occurs, it should be failure analyzed to identify the failure mechanism (the root cause of the failure). Throughout the test, means should be provided so that intermittent failures can be detected. For example, on electronic equipment, power should be continuously applied to the input circuitry, and the output circuitry should be monitored to detect intermittence.

When a failure occurs, accurate data need to be recorded on the circumstances related to the failure. It is not enough to know that the product failed. You need to determine when it failed and under what circumstances. In some cases, life testing could continue for an extended duration that provides little or no additional information. As a result, life testing is often limited to two times the projected life expectancy of the product under test.

Products that successfully pass the life test should be recharacterized and compared to their initial characterization readings to identify defects and to measure drifts in performance characteristics. Frequently, drifts in key measurements are warnings of potential failures and warrant additional study and failure analysis.

- Task 6—Perform safety factor analysis
In many cases, products will be tested at levels well above their projected customer usage requirements to measure the safety factor designed into the product. These tests typically push the product to failure. Example: raising the hi-pot voltage or electromagnetic interference noise level to the point that the unit malfunctions. These tests can provide you with excellent insight into your competitor's design strategy. The RET should also examine the competitor's product to determine any and all unique features designed into the product to provide safety protection to the customer/consumer, even if the customer is misusing the product.

- Task 7—Perform environmental tests
Environmental tests are designed to define how the product functions under extreme external conditions. Typically, these tests are performed at 10–20% higher stress levels than the actual external environment that the item is required to operate under. Typical environmental tests are temperature, vibration, shock, input voltage variation, humidity, static discharge, etc.

The environmental conditions can be applied one at a time or in combination. Maximum stress can be realized when they are applied in combination and rotated from one environmental extreme to another. For example, computers are often tested at maximum humidity while cycling them from high temperature to low temperature, and subjecting them to random frequency and magnitude vibration. Glassware can be cycled from a tub of boiling water to a tub of ice water, while being shock tested at the same time.

It is always best to have voltage applied during the environmental testing of electronic components. Circuitry should be carefully monitored to ensure that intermittent failures do not occur. If failures do occur, information needs to be collected on the exact time of failure and the environmental conditions the product is being subjected to at the time of failure. All failures should undergo a thorough failure analysis to determine their failure mechanism. Often, life testing and environmental testing are combined to reduce sample size and to increase potential failure rates.

Products that successfully complete the environmental tests should be recharacterized and compared to their initial readings to identify drifts. Frequently, drifts in key measurements are warnings of potential failures. These products are excellent candidates for further evaluation or failure analysis.

- Task 8—Compare performance results
As the three different types of tests are completed, the results of the tests and the failure analysis should be added to the database. The control sample performance should now be compared to the competitor's product's performance. All areas where the competitor's product outperforms your product should be considered improvement opportunities and be added to the root cause/corrective action database. The failure analysis activity should provide you with much of the root cause data needed to develop future corrective action plans. The product disassembly analysis will also help the RET identify why the competitor's product outperforms the RET's product.

- Task 9—Perform a product disassembly analysis
There is a great deal that an organization can learn from understanding how their competitors manufacture their products. One of the best ways to accomplish this is by disassembling competitors' products and comparing the product design, assembly methods, and each component part to your product. Typical things that reverse engineering activities can reveal are
  - Number of different parts required to accomplish a specific function
  - Level of standardization of parts used by the competitor
  - Suppliers used by the competitor
  - Actual tolerance variations
  - Assembly methods
  - Lubrications used
  - Materials used
  - Ease of repair
  - Etc.

I have seen rows of engines from each of the organization's competitors disassembled and laid out in a large design laboratory. The rows were laid out north to south, showing how the engine came apart down to the component level. If you viewed the area from east to west, each row would contain the equivalent part from each of the competitors. For example, one row would contain the organization's and its competitors' pistons, laid out for easy comparison.

Often, samples that have completed life tests are included in disassembly evaluation to measure how much wear the component parts have as a result of the life test. These measurements will often allow the organization to predict when the item will fail. These data also provide meaningful improvement opportunities.

A well-defined disassembly process needs to be developed and documented by the RET. It is always best to train the personnel who will be doing the disassembly activities by having them disassemble and reassemble a number of your own products. It is important to realize that products are designed to be assembled, not for ease of disassembly. Disassembling a product without damaging it is a real art and requires highly skilled individuals. Great care must be used not to damage the item as a result of the disassembly process.

Products are designed today to facilitate easy, fast repair. Throwaway assemblies are often used because it costs too much and requires too much skill to repair the item at the component level. If a customer has to pay $56 an hour to a repair person who takes 2 hours to diagnose a defective resistor and replace it in an assembly that only costs $22, you are not providing good customer service.

The personnel used to disassemble the product need to be highly skilled technicians who have a great deal of creativity and understanding of the function of each component. In addition, part of the disassembly team has to have in-depth knowledge of the process you use to produce the product. Little things are critical here. The difference between using a flat washer and a lock washer can be critical.

A major part of the disassembly analysis is dedicated to defining the difference in the cost to correct similar problems in the competitors' products versus your products. The disassembly sample should provide adequate parts for destructive testing of component products. Example: measuring plating thickness, hardness testing, materials analysis, etc.

Adequate space must be set aside to do the disassembly. In most cases, this space must be kept very clean because the component parts often have oil and lubricants on them that attract dirt. We like to use at least a class 1000 cleanroom. One of the mistakes made by organizations that are just starting their product benchmarking activities is to underestimate the space required to lay out the disassembled parts and the length of time the space will be required.

Once you have trained personnel, a disassembly area, and characterized products, you can start the disassembly process. A key person on the disassembly team is an experienced video camera operator who has good video equipment and appropriate lighting. It is extremely important to carefully record the entire process so that no detail will be missed. It is also very valuable to have a disassembly record that will help the disassembly team reassemble the competitor's products.

It is advisable to disassemble two products in parallel with each other: one of your own products and one of the competitor's products. The disassembly team should divide the work into small tasks (e.g., pull the engine block). You should then perform the disassembly task on your own product first and repeat the task on the competitor's product, comparing the differences. Care should be exercised to keep excellent records. Typical things that should be recorded are

- Number and types of different tools used
- Can it be done with standard tools?
- All clearances and adjustment measurements (e.g., spring tension, timing, torque requirements to unloosen screws, etc.)
- Amount of lubricant
- Parts suppliers
- General workmanship comments
- Etc.

This process is repeated until the products are disassembled to the desired level. Once the product is disassembled, the key individual parts are characterized. Here again, variable data is extremely important. After the component parts have been characterized, the disassembly team should review its disassembly log and the disassembly video. The team will then prepare an assembly procedure for the competitor's item. This assembly procedure will be used to reassemble the competitor's item. The disassembly team will use the normal manufacturing procedures to reassemble its own product. The disassembly team should follow its version of the competitor's assembly process as close as possible. When it is necessary to deviate from the documented assembly procedures, the procedures should be changed so that they reflect exactly how the item was reassembled. It is important to note that it is not practical for fixtures to be made to support this assembly process. As a result, some differences can occur. When the products are reassembled, they should be recharacterized to ensure that the simulated assembly process provides compatible products.

- Task 10—Compare product design and production methods
  The disassembly team has collected a great deal of data and opinions during the preceding activities. These data need to be analyzed on an ongoing basis during the disassembly and assembly processes. Key differences between the RET's and the competitor's item need to be identified. Differences will exist, but that does not mean that the competitor's product is better. The competitor's drive gear may be made of a different material that is harder than yours, but is that good or bad? Disassembly analysis could reveal that this gear wore much less than the benchmarking organization's gear during its life cycle, but the other material costs significantly more. It is up to the disassembly team to define the differences between the products and list the pros and cons of these differences. This information should be entered into the database.

- Task 11—Define competitor's competitive advantage
  The data collected during the characterization, life, safety factor, and environmental tests were used to define improvement opportunities based on a comparison of your competitor's products and your product's performance. The disassembly analysis provides you with a good understanding of the product design and production methods. The disassembly analysis process can also provide you with additional improvement opportunities and insight into why the competitor's products perform better than your products during the test phase.

  Now these two databases need to be analyzed to define where the competitor's competitive advantages are. The RET then needs to review each improvement opportunity to determine if it provides the competitor with a true competitive advantage. To put it another way, the RET needs to evaluate each opportunity to determine if making the change is truly value added to the stakeholders. The RET should ask itself: would a potential change decrease cost and/or increase customer satisfaction? It is easy for the RET to want to pursue changes that make your products perform better than your competitors, but that does not have a positive impact on the customer or the organization. These types of changes are a waste of time, effort, and money.

Positive-impact areas for the customer are reduced cost, increased features, improved quality, and ease of operation. Positive-impact areas to the organization are increased market share or a decrease in the resources required to produce the products, resulting in a bigger profit margin. Do not get carried away with improvements for improvement's sake. All improvements cost money. Adding performance that will never be used is wasteful.

SUMMARY

Many people feel that reverse engineering is unlawful or, at least, unethical. Neither is true. If you are not doing reverse engineering, you are putting your organization at a severe competitive disadvantage because your competition is probably testing and disassembling your latest products as you read this paragraph. There is nothing unethical about understanding your competition's products, as long as you do not copy patented parts of their product. Gaining knowledge about who your suppliers are, how their products are designed, and how well their products perform can be a very important part of helping the product engineering function make the best decisions related to your next-generation products.

Combining reverse engineering with a good research/data collection system is imperative. Information that is in the public domain provides an effective tool for projecting what performance breakthroughs are in the mill and approximately when they will be made available to the general public. These are essential inputs to the product's reliability specification and the product's engineering design considerations.

EXAMPLES

Examples are included in the part of this chapter entitled "How to Use the Tool."

SOFTWARE

We have no specific software recommendations.

REFERENCES

Harrington, H.J. *The Complete Benchmarking Implementation Guide*. New York: McGraw-Hill, 1996.

Harrington, H.J. and Harrington, J.S. *High Performance Benchmarking—20 Steps to Success*. New York: McGraw-Hill, 1996.

15

Robust Design

Stuart Burge

CONTENTS

This chapter has been produced in association with Paul Goodstadt, ACIB, FRSA, FTQMC, director of development at the Total Quality Management College, Manchester, United Kingdom.

DEFINITION

Robust design is more than a tool; it is complete methodology that can be used in the design of systems (products or processes) to ensure that they perform consistently in the hands of the customer. It comprises a process and tool kit that allows the designer to assess the impact of variation that the system is likely to experience in use, and if necessary redesign the system if it is found to be sensitive.

USER

This tool can be used by individuals, but its best use is with a group of four to eight people. Cross-functional teams usually yield the best results from this activity.

OFTEN USED IN THE FOLLOWING PHASES OF THE INNOVATIVE PROCESS

The following are the seven phases of the innovative cycle. An X after the phase name indicates that the tool/methodology is used during that specific phase.

- Creation phase X
- Value proposition phase
- Resourcing phase
- Documentation phase
- Production phase X
- Sales/delivery phase
- Performance analysis phase

TOOL ACTIVITY BY PHASE

- Creation phase—During this phase, the tool is used to stimulate creative thinking and to evaluate solutions to ensure the product or process design will adequately fulfill the customers'/consumers' requirements at a reasonable cost level.
- Production phase—During this phase, the tool provides information on allowable manufacturing variation of the design parameters that can be used to establish Control Charts and process parameter settings via Quality Function Deployment.

HOW TO USE THE TOOL

The creator of robust design is Dr. Genichi Taguchi who, in the 1950s, worked for the Electrical Communications Laboratory (ECL) of the Nippon Telegraph and Telephone Corporation. During his 12 years there, he developed his ideas and understanding of robust design. One critical insight was the recognition that small changes (variation) in a design parameter can significantly affect the performance. Taguchi also realized that the traditional approach to handling this variation through use tolerances and specification limits can be misleading. Taguchi was able to put his understanding on a sound theoretical basis through the concept of the *loss function*. But his stroke of genius was the understanding that the inherent nonlinear nature of most systems can be exploited to find design solutions that are *robust* against the variation they are likely to experience in use. Rather than tightening tolerances when faced with a sensitive design, the system should be redesigned to find an alternative solution that was insensitive. He distilled this into an approach he called *parameter design.*

Robust design is now an established methodology for most serious engineering companies, and has made a major contribution to product quality. It is one of the reasons your motorcar starts first time every day and your mini-system has such good sound quality. Taguchi was interested in product design, but it is clear now that robust design is applicable to any type of system, such as process- or service-intensive systems. The same principle can be applied to the service quality in a hotel, restaurant, or bank.

Taguchi's Loss Function

When working at ECL, Taguchi noticed that apparently minor variations in a system design parameter can have a significant effect on overall system performance. He also recognized that the *traditional* approach to handling such sensitivities by setting tight tolerances is flawed.

To explain Taguchi's philosophy, it is worthwhile considering a simple example. Imagine that a lawn-mower designer has selected a rotating blade, like that shown in Figure 15.1, as the mechanism for cutting grass.

FIGURE 15.1
Lawn mower blade.

One of the responsibilities of the designer is to determine numerical values for the blade design parameters, such as

- Blade length
- Blade thickness
- Blade width
- Blade shape
- Blade material
- Etc.

These numerical values are called the *target* or *nominal* values, and represent the ideal values that typically the designer has calculated to give the required performance. Most designers, however, know that these target values will not be achieved in practice because of one or more causes of variation that include

- Environmental, for example, temperature, pressure, humidity etc.
- Manufacturing: No two components can be made the same.
- Wear: Over time, wear causes the values to change.

To *cope* with this expected variation, designers will set limits (called specification limits or tolerances), as shown in Figure 15.2.

The purpose of specification limits, or tolerances, is to identify when an actual blade is so far from its intended design target value that the product performance is unacceptable and potentially leading to customer complaints. Taguchi noted that these specification limits were often chosen from experience and judgment, or simply copied from the previous design—there was little science in establishing the *best* limits.

From a quality viewpoint, items outside the specification limits are to be rejected. On identifying a reject, there are three courses of action:

1. Scrap the reject and make a replacement. The money for this replacement will come from the company's profit.

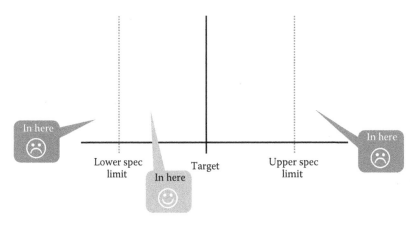

FIGURE 15.2
Use of specification limits (tolerances) to manage variation.

2. Rework the reject to bring it inside the limits. The money for this rework will come from the company's profit.
3. Agree on a concession, a relaxation of the specification limits, on a case-by-case basis. The money for this procedural work will come from the company's profit.

The important point to note is that if an item is outside the specification limits, it is going to cost and the money will come from this year's profit. However, an item inside the limits will not require the expenditure of money.

This cost of variation is called the quality loss and, based on the scenario described above, results in the economic model shown in Figure 15.3. Looking at this figure, you will note that when the variation in the design parameter reaches the specification limit, there is a step change in cost. It may not be equal at either side of the target, but inside the limits the loss is zero.

Taguchi was unhappy with this quality loss model, because it inferred that an item exactly on the designer's target would have a zero loss, as would one just inside the specification limits. Remember that these specification limits are often determined by *engineering judgment* or what was on the previous design. Taguchi argued that an item near the specification limit could not be the same as one exactly on target. The cost model was wrong and a better model was needed.

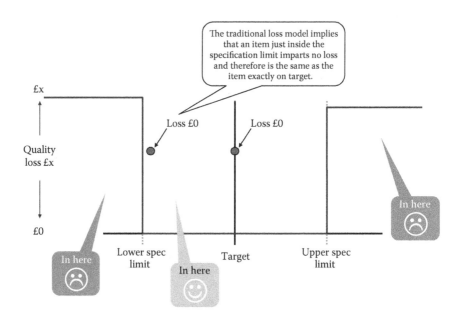

FIGURE 15.3
Traditional quality loss model.

In deriving a new cost model, Taguchi's key insight was that a loss would always be incurred when a design parameter (denoted by y) deviates from its target value (denoted by m) regardless of how small the deviation is. Taguchi argued the quality loss is zero only when $y = m$ and defined a loss function as $L(y)$ where

$$L(y) = L(m + \{y - m\})$$

which could be expanded by Taylor's series

$$f(x + h) = f(x) + hf'(x) + (h^2/2!)f''(x) + (h^3/3!)f'''(x) + \ldots$$

by making $x = m$ and $h = \{y - m\}$. Thus

$$L(y) = L(m) + L'(m)\{y - m\} + L''(m)\{y - m\}^2/2! + \ldots$$

Returning to Taguchi's premise, when $y = m$, the loss is zero—that is, $L(y) = 0$, and since $\{y - m\}$ will be small, terms with power higher than two can be ignored, resulting in

$$L(y) = L''(m)\{y - m\}^2/2! = k(y - m)^2$$

The term $(y - m)$ represents the deviation from the target value and so the loss A due to a deviation $\Delta = (y - m)$ is

$$A = k\Delta^2$$

In other words, the loss is quadratic, as shown in Figure 15.4.

The profound insight that comes from Figure 15.4 is that even when we are inside the *specification limits*, there will be a quality loss measured in pounds, yen, or dollars. Specification limits have no economic value. Dependent on which side of the target you are, sometimes the company loses, and sometimes the customer will lose. Taguchi summed this up by stating that any product that does not meet its target specification will impart a loss to society. He also said, "A company that ships a product with a high quality loss is worse than a thief!"

Potentially, every design parameter in a system will have its own characteristic loss function, but the *degree of loss* is dependent on the constant k. Thus, for different k values, there will be a different *shape* to the loss function (still quadratic) that reflects the sensitivity of the chosen design point. Some will be sensitive, some will not—some will be robust to variation, as shown in Figure 15.5.

Figure 15.5 shows two extreme cases. The first is a sensitive, or unrobust design, where variations in the particular design parameter around the

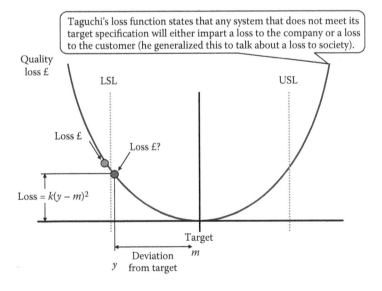

FIGURE 15.4
Taguchi's loss function.

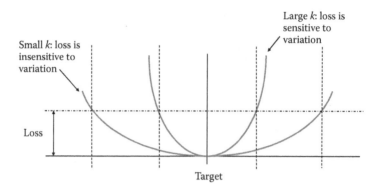

FIGURE 15.5
Sensitive and insensitive loss functions.

target value will cause significant quality losses. Faced with the knowledge of such a situation, the typical reaction of the designer is to tighten the specification limits. There is no doubt this will work, the system will work as expected in the hands of the user, but the tightening of the tolerances adds cost that either must be borne by the user or the producer. Someone is going to lose out.

The second extreme is the flatter loss function. Here, the performance will only degrade when the variation is large. If variation occurs over the expected range, then the user will experience very little degradation in performance. This is a robust design, where performance is maintained at the desired level over the expected range of operating conditions.

In other words, having a loss function that is insensitive to variation is desirable because the end user, the customer, will experience consistent performance. It was this epiphany that led Taguchi to his robust design process. Clearly, it is the responsibility of the designer to determine the loss function, and if it is sensitive, not to react by setting tighter specification limits but to redesign to a new design point where it is insensitive to variation.

To reinforce this critical message, consider the system shown in Figure 15.6. This particular system has only two design parameters x_1 and x_2. When x_1 and x_2 are changed, they will put the design on another point on the *mountain*. The customer requires a *certain* performance, which is represented by the plane that cuts through the mountain. Anywhere above the *plane* is acceptable; anywhere below is not.

Faced with such a situation and the pressure to deliver *excellent* performance, many a designer would aim for the peak of the mountain. This

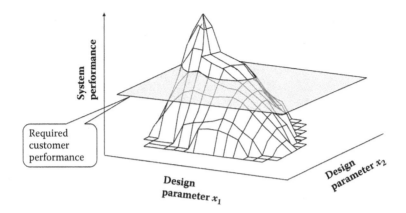

FIGURE 15.6
System design with two design parameters.

is simply because it is *world-class* performance, it cannot be bettered. Indeed, having found the peak, the designer could work *backwards* to find appropriate values for x_1 and x_2, as shown in Figure 15.7.

Theoretically, the situation shown in Figure 15.7 is the best possible design since it gives the best possible performance. However, it is entirely reliant on being able to

- Achieve the target values for x_1 and x_2
- Maintain the target values for x_1 and x_2

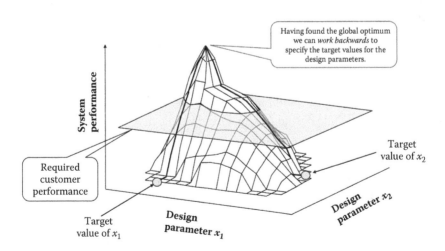

FIGURE 15.7
Designing the *best* possible system.

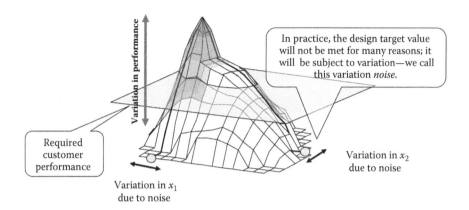

FIGURE 15.8
Impact of noise on system performance.

Neither of these is possible in the real world because of variation caused by environmental factors, manufacturing variation, and wear. In other words, the target values will vary considerably and we "will fall off the mountain peak," as shown in Figure 15.8. Taguchi called this real-world variation *noise.*

Because of the noise, the variation in system performance experienced by the customer is huge. One day the performance will be world-class, the next it will be awful. Customers hate inconsistent performance—they like consistent performance. This is a sensitive situation, and the loss function at this peak is the steep one in Figure 15.5.

Returning to the earlier argument around Figure 15.5, having found yourself at the peak and discovering the extreme sensitivity to noise, one solution is to set *tighter* tolerances around the design parameters. There is no doubt this will work, as shown in Figure 15.9, but at a greater cost.

By now, however, you will have realized there is another solution, which is to move the whole design to new target values that put it on the plateau. By placing the design on the plateau, we find the following:

- We achieve the customers' performance requirements.
- We are insensitive to the likely variation (the noise).

We will, of course, not achieve the *best* possible performance, but it does exceed their requirements consistently. This plateau, shown in Figure 15.10, is a *robust optimum.*

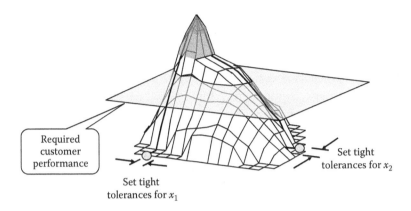

FIGURE 15.9
Tightening tolerances—an expensive solution.

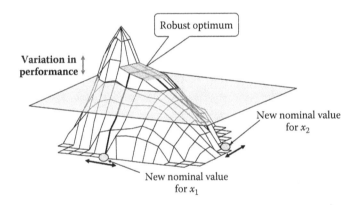

FIGURE 15.10
Place to be: the robust optimum.

The beauty of a robust optimum is that, although we do not achieve the best possible performance, we do consistently exceed our customer's expectations. Customers like consistent performance. When on the plateau, the loss function is the flatter one in Figure 15.5.

Taguchi's Robust Design Approach

Taguchi codified the ideas expressed above into the three-stage process:

1. System design
2. Parameter design
3. Tolerance design

While this process is broadly correct because it is at such a high level, it is possible to add more detail as to how to perform robust design. This more detailed process is shown in Figure 15.11.

Engineer system is about creating a system design concept that satisfies the customer's requirements. Taguchi talked little about how to do this, but it is a critical step. Optimizing a *bad* system concept design is nugatory work. To ensure the best possible solution, systems engineering (Blanchard and Fabrycky, 2010) must be used, as this will demand the full

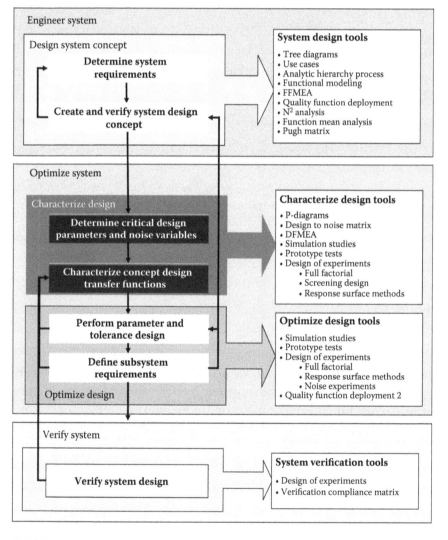

FIGURE 15.11
Whole robust design process.

understanding of the problem defined by the customer's requirements, the exploration of candidate solutions ultimately leading to the selection of the *best* system concept solution.

Optimize system comprises two activities. The first is to *characterize design*. This is, in simple terms, finding out what the mountain (range) looks like. I want to know, either explicitly or implicitly, the mathematical relationship between my design parameters, noise factors, user commands, and the critical system performance measures that are important to the customer. Taguchi reduced any robust design problem down to the diagram shown in Figure 15.12.

Taguchi argued that there are three types of factors that determine the output of a system:

- *Signal factors M* (or user commands) are parameters set by the user to set or command the intended value for the output of the system. So, for example, on a lawn mower, the user may have a throttle lever to change the speed of rotation of the cutter blade.
- *Control factors Z* (or design parameters) are parameters that can be freely specified by the designer. In fact, it is the designer's responsibility to determine the best values for these parameters. For example, the lawn-mower design may decide the blade length is 520 mm.
- *Noise factors N* are factors or parameters that cannot be, or chosen not to be, controlled by the designer. They cause the output Y to deviate from the target specified by the signal factor M.

Taguchi said that the job of a designer is to determine the target values for the design parameters (control factors) such that the system output achieves the user-commanded level irrespective of the noise factors. Pictorially, as shown in Figure 15.13, we want the design on a plateau and

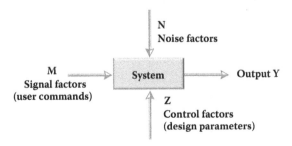

FIGURE 15.12
Universal robust design problem.

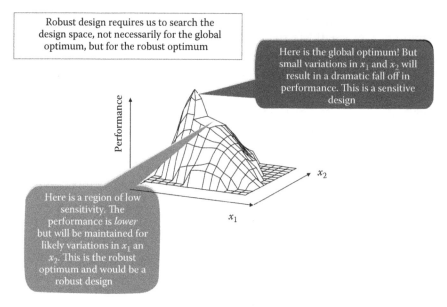

FIGURE 15.13
Robust design task.

not on a peak. It is therefore necessary to search the mountain range to find the plateaus and avoid the peaks. This searching of the design space—the mountain range—is what Taguchi called parameter design.

Figure 15.13 shows a very simple system with only two design parameters where it is possible to visualize the relationship between the design parameters and the system output. Real systems have much more than two design parameters; they cannot be visualized in the way Figure 15.13 shows. Effectively, the robust design task is akin to a blind person searching a mountain range to find the plateaus and avoid the peaks.

Given a concept design from systems engineering, the first step in robust design is to determine all the design parameters, noise factors, and signal factors that relate to a particular system output that, in turn, relates to customer satisfaction. Here, the tool of choice is a form of Figure 15.12 called a *parameter diagram* or *P-diagram*. Figure 15.14 shows an example P-diagram for the grass-cutting system of a lawn mower.

A P-diagram can help a designer (or design team) identify all the possible factors that could have a role to play in delivering good cut performance. Figure 15.14 is partially complete but shows the overall nature of the *design problem*. It is not uncommon to have 50–100 noise factors and 20–50 design parameters. To be able to construct the *mountain range*

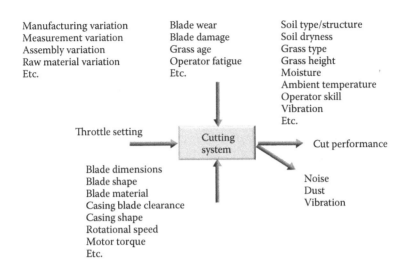

Manufacturing variation
Measurement variation
Assembly variation
Raw material variation
Etc.

Blade wear
Blade damage
Grass age
Operator fatigue
Etc.

Soil type/structure
Soil dryness
Grass type
Grass height
Moisture
Ambient temperature
Operator skill
Vibration
Etc.

Throttle setting

Cutting system

Cut performance

Blade dimensions
Blade shape
Blade material
Casing blade clearance
Casing shape
Rotational speed
Motor torque
Etc.

Noise
Dust
Vibration

FIGURE 15.14
Partial P-diagram for the grass-cutting system of a lawn mower.

means that we need to determine the *transfer function* between the input factors and the system output. In other words, we need to know mathematically the relationship

$$Y = f(Z, N, M)$$

We do not have to know this mathematical expression explicitly (as an actual formula); we can undertake robust design if we know the relationships implicitly usually through a simulation model of the system. For engineered systems, we build finite element models to determine the stress level, to ensure the design is strong enough. For process-based systems, we can build process simulation models to investigate bottlenecks and stock levels.

It is clear, however, that creating a transfer function with several hundred terms is not easy, and in fact we do not do this. Having identified all possible factors, we now determine which of these are critical. A useful tool here is the design to noise matrix, which is a simple tool to allow a designer or design team to assess which factors (design and noise) are most likely to influence the robustness of a nominal design. It is used to provide a quantitative assessment of the influence that both design parameters and noise factors have on the outputs of a system to determine which are critical. Typically, we aim to reduce the number of design and noise factors to between 10 and 20. Any more and the problem becomes intractable.

Even with a reduced number of design and noise factors, to fully map out the mountain range (while not impossible) may take too long. What Taguchi wanted is an efficient way of searching the mountain range—the design space. Fortunately, while he worked at Electrical Communications Laboratory, he was a visiting professor at the Indian Statistical Institute, where he worked with Ronald Fisher and C.R. Rao. Fisher and Rao were responsible for the development of the design of experiments (DoE) and, in particular, the orthogonal array.

DoE is the name given to a collection of experimental approaches that explore the effect changing multiple system parameters have on the output of a system. Its power stems from its ability to quantify not only the effects of the individual parameters (the main effects) but also the interactions between parameters. Its origin can be traced back to the pioneering work of Sir Ronald Fisher (1890–1962) when working at the Royal Agricultural College Rothamsted. Very much the preserve of statisticians for several decades, it was Taguchi who turned a relatively complex methodology into a practical everyday tool. The advent of statistical software packages like Minitab (www.minitab.com) have now made its use even more widespread.

DoE is one of the most powerful of experimental approaches. It is, however, counterintuitive to the uninitiated. In simple terms, a DoE provides a method for designing and analyzing results from an experiment to understand and quantify the effects that multiple factors have on the outputs of a system. Usually, when faced with the scenario where the output of a system could be affected by numerous factors, most engineers would resort to changing just one factor at a time. Changing two or more would be seen as foolhardy because "it would not be possible to separate out the contribution of each factor." A DoE, however, is a system of experimental trials where several factors are changed for each trial in a way that will allow for the separation of the various contributions from each factor. It provides a powerful way of searching a design space.

What Taguchi did was to take the work of Fisher and Rao to develop a number of design arrays like that shown in Table 15.1. Table 15.1 shows an L6 design array where it is possible to explore six control factors (design parameters) in eight trials at two experimental levels (represented by 1 and 2). In the early days of using Taguchi's approach, organizations would build physical prototypes where the design parameters considered being

TABLE 15.1

L6 Design Array

Trial	Z_1	Z_2	Z_3	Z_4	Z_5	Z_6
1	1	1	1	2	1	1
2	1	1	2	1	2	2
3	1	2	1	2	2	2
4	1	2	2	1	1	1
5	2	1	1	2	1	2
6	2	1	2	1	2	1
7	2	2	1	2	2	1
8	2	2	2	1	1	2

important could be adjusted according to the design array. If nonlinear behavior were suspected, then three or four experimental levels would be used. The experimental trials would be run, and the result analyzed to find out which design parameters are important.

The analysis of the experimental results can be plotted in various ways, one of which is a cube plot. Figure 15.15 shows a cube plot for the results from an L6 design. The length, width, and depth for each cube relate to

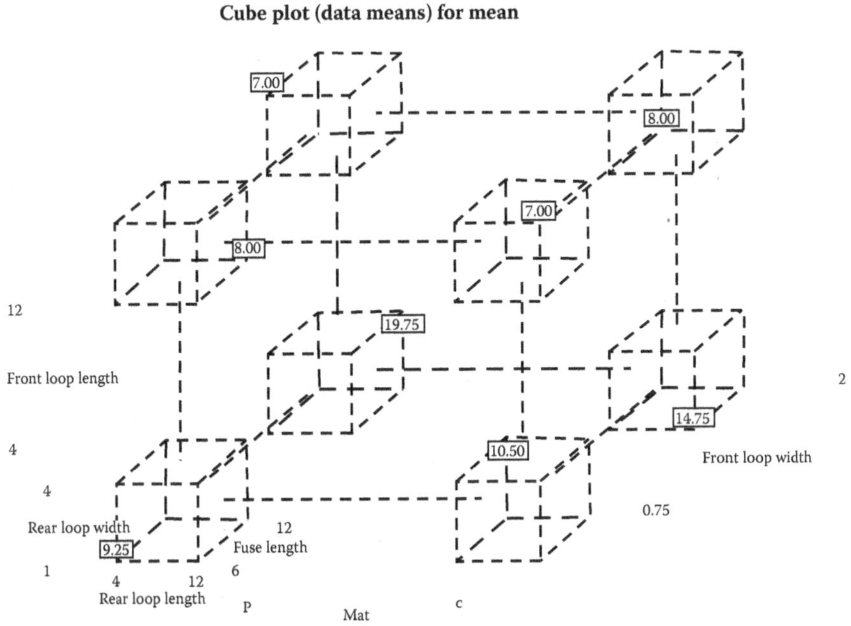

FIGURE 15.15

Cube plot of results from an L6 design of experiments.

three of the design parameters. Right and left cubes relate to the fourth, front and back to the fifth, and up and down to the sixth. Note that the eight experimental results are recorded on the cube of cubes and provide insight to the design space.

If we look at the top four cubes in Figure 15.15, the results are 7.00, 8.00, 7.00, and 8.00. This could be a plateau as the results are quite stable, whereas the lower level of cubes show considerable variation and present a maximum value of 15.75 on the rear left cube. This could reflect a peak in the design space.

These days, it may not be necessary to conduct actual physical experiments. Instead, computer-based models that simulate the behavior of a system can be used. Such models combined with Monte Carlo* approaches can identify the plateaus associated with a robust design. Here, the DoE can be used to decide on the simulation runs necessary to efficiently explore the design space. A numerical example is provided for illustration.

EXAMPLES

Examples are displayed throughout the portion of this chapter entitled "How to Use the Tool."

SUMMARY

Robust design is a complete methodology for designing systems that are insensitive to the variation they are likely to experience in use. The consequence is that such systems will perform consistently in the hands of the user. Robust design was developed by Dr. Genichi Taguchi who adopted a systems view of the problem that led to profound understanding of the impact of variation through Taguchi's loss function. Moreover, Taguchi realized that

---

* Monte Carlo is a mathematical approach to the modeling of the effects of variation in a system. In particular, it looks at the effect of input variation and how this gets transmitted into output variation through the system's transfer function by randomly generating values for the system inputs and calculating their associated output values. This exercise is literally repeated hundreds, if not thousands, of times to *build up a picture* of the input to output variation. While its origin was in the Manhattan Project and performed by hand calculation, modern computers and software make it an invaluable tool for the robust designer.

he could exploit the nonlinear behavior of most systems to find regions of the design space that are insensitive. To find these regions, Taguchi developed the use of DoE as an efficient way of exploring the design space.

Since robust design has received consistent attention from researchers and practitioners for years, a number of methodologies for robust design optimization have been reported in the research community. However, the majority of these existing methodologies ignore the case where the customer may tolerate and specify an upper bound on process bias. This notion also suggests a bias-specified robust design method and formulation of a nonlinear program that minimizes process variability subject to customer-specified constraints on the process bias using the ε-constrained method.

SOFTWARE

Some of the commercial software available include but are not limited to

- Minitab Statistical Software: www.minitab.com, 2014

REFERENCE

Blanchard, B.S. and Fabrycky, W.F. *Systems Engineering and Analysis*. Upper Saddle River, NJ: Prentice Hall International Series in Industrial & Systems Engineering, 2010. ISBN-13: 978-0132217354.

SUGGESTED ADDITIONAL READING

Shin, S. and Cho, B.R. *Bias-Specified Robust Design Optimization and Its Analytical Solutions*. *Computers & Industrial Engineering*, vol. 48, no. 1, pp. 129–140, 2005. Selected papers from the 31st International Conference on Computers and Industrial Engineering.
Skogestad, S. and Postlethwaite, I. *Multivariable Feedback Control—Analysis and Design*. New York: Wiley, 2005. ISBN 0-470-01168-8.

16

SCAMPER

Frank Voehl

CONTENTS

DEFINITION

SCAMPER is a tool that helps people generate ideas for new products and services by encouraging them to think about how you could improve existing ones by using each of the seven words that SCAMPER stands for, and applying it to the new product or service in order to generate additional new ideas. SCAMPER is a mnemonic that stands for:

1. Substitute
2. Combine
3. Adapt
4. Modify
5. Put to another use
6. Eliminate
7. Reverse

USER

This tool can be used by individuals or groups, but its best use is with a group of four to eight people. Cross-functional teams usually yield the best results from this activity.

OFTEN USED IN THE FOLLOWING PHASES OF THE INNOVATIVE PROCESS

The following are the seven phases of the innovative cycle. An X after the phase name indicates that the tool/methodology is used during that specific phase.

- Creation phase X
- Value proposition phase X
- Resourcing phase X
- Documentation phase X
- Production phase X
- Sales/delivery phase X
- Performance analysis phase

TOOL ACTIVITY BY PHASE

SCAMPER is a tool that can be used in all the phases of the innovation cycle when problems occur or when new approaches are being developed. It is particularly useful during the creation phase and the production phase.

HOW TO USE THE TOOL

SCAMPER is an acronym for seven thinking techniques that help those who use it to come up with untypical solutions to problems. The thinking techniques are so common to human creative behavior that it might be more accurate to call SCAMPER a mnemonic for the collection of techniques rather than a technique of its own.

SCAMPER is used when it can often be difficult to come up with new ideas or when you are trying to develop or improve a product or service. This is where creative brainstorming techniques like SCAMPER can help. This tool helps you generate ideas for new products and services by encouraging you to think about how you could improve existing ones.

You use the tool by asking questions about existing products, using each of the seven prompts:

1. Substitute
2. Combine
3. Adapt
4. Modify
5. Put to another use
6. Eliminate
7. Reverse

These questions help you come up with creative ideas for developing new products, and for improving current ones. Alex Osborn, credited by many as the originator of brainstorming, originally came up with many of the questions used in the SCAMPER technique. However, it was Bob Eberle, an education administrator and author, who organized these questions into the SCAMPER mnemonic.

SCAMPER is really easy to use. To use SCAMPER, you simply go down the list and ask questions regarding each element. Remember, not every idea you generate using SCAMPER will be viable; however, you can take good ideas and explore them further. It is important to remember that the word *products* does not only refer to physical goods. Products can also include processes, services, and even people. You can therefore adapt this technique to a wide range of situations as follows:

1. First, take an existing product or service. This could be one that you want to improve, one that you are currently having problems with, or one that you think could be a good starting point for future development.
2. Then, ask questions about the product you identified, using the SCAMPER mnemonic to guide you. Brainstorm as many questions and answers as you can. (We have included some example questions in this chapter.)
3. Also, look at the answers that you came up with. Do any stand out as viable solutions? Could you use any of them to create a new product, or develop an existing one? If any of your ideas seem viable, then you can explore them further.
4. Finally, be aware to ask SCAMPER questions.

Let us look at some of the questions you could ask for each letter of the SCAMPER mnemonic.

Substitute

- What materials or resources can you substitute or swap to improve the product?
- What other product or process could you use?
- What rules could you substitute?
- Can you use this product somewhere else, or as a substitute for something else?
- What will happen if you change your feelings or attitude toward this product?

Combine

- What would happen if you combined this product with another, to create something new?
- What if you combined purposes or objectives?
- What could you combine to maximize the uses of this product?
- How could you combine talent and resources to create a new approach to this product?

Adapt

- How could you adapt or readjust this product to serve another purpose or use?
- What else is the product like?
- Who or what could you emulate to adapt this product?
- What else is like your product?
- What other context could you put your product into?
- What other products or ideas could you use for inspiration?

Modify

- How could you change the shape, look, or feel of your product?
- What could you add to modify this product?
- What could you emphasize or highlight to create more value?
- What element of this product could you strengthen to create something new?

Put to Another Use

- Can you use this product somewhere else, perhaps in another industry?
- Who else could use this product?
- How would this product behave differently in another setting?
- Could you recycle the waste from this product to make something new?

Eliminate

- How could you streamline or simplify this product?
- What features, parts, or rules could you eliminate?
- What could you understate or tone down?
- How could you make it smaller, faster, lighter, or more fun?
- What would happen if you took away part of this product? What would you have in its place?

Reverse

- What would happen if you reversed this process or sequenced things differently?
- What if you try to do the exact opposite of what you are trying to do now?
- What components could you substitute to change the order of this product?
- What roles could you reverse or swap?
- How could you reorganize this product?

Tips

- *Tip 1*

  Some ideas that you generate using the tool may be impractical or may not suit your circumstances. Do not worry about this—the aim is to generate as many ideas as you can.
- *Tip 2*

  To get the greatest benefit, use SCAMPER alongside other creative brainstorming and lateral thinking techniques.

For example, a bicycle has the following components: frame, pedals, drive sprocket, chain, brakes, tires, handlebars, and so forth. By applying SCAMPER, the following ideas for improving the bicycle can be generated:

- Much lighter weight frames based on new materials
- Pedal grips that strap to secure feet better
- Stronger chains with special clamps for easier changing
- Improved derailleur gears for rear sprocket
- Racing handlebars for more ergo dynamic racing position
- New rear wheel materials to replace spokes

Advantages

1. By using the SCAMPER method, an individual or a group may be spurred into generating new ideas by simply evaluating an existing one.
2. This process can result in vast improvements being made to both products that may exist already and to product ideas that are still in their infant stages.

3. Where an idea may have encountered a development obstacle, the SCAMPER method may prove to be a systematic approach to overcoming that obstacle, allowing new ideas to be generated and an improved product to come to fruition.
4. The SCAMPER process is also largely used in regard to encouraging the creative process in the minds of employees and school children alike, influencing the generation of new ideas without placing boundaries on where they come from.
5. Some people have trouble with the development of new ideas when they have not been provided much of a creative influence, and the SCAMPER method can be used by educators to influence the generation of creativity in participants by using the process to promote creative thinking.
6. This process has also been largely proven to promote constructive ideation and problem-solving abilities in children by engaging their minds to think around obstacles in order to overcome them.

Disadvantages

1. SCAMPER is a tool requiring the right environment.
2. One weakness the SCAMPER method may be prone to is that the correct environment is needed for the method to be proven effective.
3. For the process to truly influence creativity, both children and employees must be in an environment that truly promotes and encourages new ideas, regardless of whether or not the ideas may be considered useful.
4. If this environment has been proven to be nonexistent, the process will be proven ineffective because of the lack of constructive reinforcement throughout the process.

SUMMARY

SCAMPER is a technique you can use to spark your creativity and help you overcome any challenge you may be facing. In essence, SCAMPER is a general-purpose checklist with idea-spurring questions—which is both easy to use and surprisingly powerful. It was created by Bob Eberle in the early 1970s, and it definitely stood the test of time. To use the SCAMPER technique, first state the problem you would like to solve or the idea you would like to develop. It can be anything: a challenge in your personal life or business, or maybe a product, service, or process you want to improve. After pinpointing the challenge, it is then a matter of asking questions about it using the SCAMPER checklist to guide you.

EXAMPLE

For details about the following case study, see http://forthewater.blog .com/2013/02/02/scamper-case-study/.

Jonathan was just an ordinary Singaporean when he found out that in Africa, there are people such as Naziah, who were suffering from water scarcity. Jonathan, being the helpful and kind person he is, feels the need to help people like Naziah by inventing a product that can help reuse used water. As time passed, Jonathan and his team completed the sketch of the product but just as they were about to build the product, they saw that there were some problems with their design. The following are some of the challenges faced by Jonathan and his team and how they used the SCAMPER method to solve them.

> *Substitute:* The initial idea was to install a filter into a portable water bottle so that the unclean water gets filtered as it is drunk. However, they realized that the filter was way over the budget and it was also too big for the design. Therefore, they substituted the filter for materials used in a filter, for example, charcoal, gravel, sand, and stones.
>
> *Combine:* The opening where the filtered water comes out is inconvenient toward the user as the opening is at the bottom where it is hard to retrieve the water. Thus, they combined the idea of the tap to the product by attaching it to the side so that the user can retrieve water easily.
>
> *Adapt:* The product could not stand on its own and was hard to use as the user had to hold the product the entire time they were using it. It was tedious and inconvenient. They then adapted to Naziah's environment by putting a stand on the bottom of the product such that the user will not need to carry it around.

Modify: The shape of the product was not suitable as the cylindrical shape was not able to equalize the amount of sand, gravel, and charcoal. The structure of the product did not allow the user to change the sand, gravel, and charcoal when they were dirty. They modified the shape of the product into a rectangle, so that the filters can filter the water more effectively. Also, they placed the sand, gravel, and charcoal onto plates. The plates can be slotted into a gap in the product. Therefore, when the filters need to be changed, the users have easy access to the filters and can change or wash it with ease.

Magnify: The product was too small and the filters were too big for the product. They magnified the product such that more water can be filtered and the filter will be able to fit in the product.

Put to other uses: They found out that the filtered water was not potable. Instead, they came up with a solution where they decided to recycle the water instead of drinking it.

Eliminate: They realized that instead of purifying water, charcoal was making it more harmful to drink. Therefore, they eliminated the charcoal, allowing the water to be purified more effectively.

Rearrange: The sequence of how the layers were placed was not able to filter water effectively. They rearranged the layers of filters and added some new ones so that the water can be filtered more effectively.

SOFTWARE

SCAMPER MindMap Software—The latest addition to the Mind Mapping Insider resources is the world's largest SCAMPER mind map, with more than 200 questions, phrases, and words that have been selected to help you with creative challenges.

- MindManager
- MindView
- iMindMap
- MindGenius
- XMind
- ConceptDraw MINDMAP: http://mindmappingsoftwareblog.com/ultimate-scamper-mind-map/

SUGGESTED ADDITIONAL READING

Brassard, M. *The Memory Jogger Plus.* Milwaukee, WI: ASQ Quality Press, 1989.

Michalko, M. *Thinkertoys: A Handbook of Creative-Thinking Techniques.* New York: Crown Publishing/Random House, 2001, 1991.

Mizuno, S., ed. *Management for Quality Improvement: The 7 New QC Tools.* Portland, OR: Productivity Press, 1988.

SCAMPER Random Testing Tool. Available at http://litemind.com/scamper-tool; also see the SCAMPER Mindmap at this site location.

17

Simulations

Douglas Nelson and Frank Voehl

CONTENTS

We're presently in the midst of a third intellectual revolution. The first came with Newton: the planets obey physical laws. The second came with Darwin: biology obeys genetic laws. In today's third revolution, we're coming to realize that even minds and societies emerge from interacting laws that can be regarded as computations. Everything is a computation.

Rudy Rucker

Torawarenai sunao-na kokoro, which means *Mind that does not stick.*

Matsushita
Founder, Matsushita Electric Industrial

DEFINITION

Simulation as used in innovation is the representation of the behavior or characteristics of one system through the use of another system, especially a computer program designed for the purpose. As such, it is both a strategy and a category of tools—and is often coupled with computer-aided innovation (CAI), which is an emerging simulation domain in the array of computer-aided technologies. CAI has been growing as a response to greater industry innovation demands for more reliability in new products.

USER

This tool can be used by individuals or by groups of four to eight people. Cross-functional teams often yield the best results from this activity when used for mew product or process design.

OFTEN USED IN THE FOLLOWING PHASES OF THE INNOVATIVE PROCESS

The following are the seven phases of the innovative cycle. An X after the phase name indicates that the tool/methodology is used during that specific phase:

- Creation phase X
- Value proposition phase
- Resourcing phase
- Documentation phase X
- Production phase X
- Sales/delivery phase X
- Performance analysis phase X

TOOL ACTIVITY BY PHASE

Simulation may be used throughout the innovation or ideation process. The benefits of simulation outlined in this chapter include the rapid provision of insights into outcomes of real-world scenarios at a minimal cost, at high levels of reliability when compared with other methods of analysis.

BACKGROUND

When Konosuto Matsushita called innovation *the mind that does not stick*, he was well aware of the management theory debate over the past 20 years that focused on companies being able to gain advantage when their organization and people are able to rapidly innovate and respond to change, and to do so at a rate faster than their competitors. To achieve this organizational state of the nonsticking mind, significant importance is being given to managers of innovation* developing skills such as being (a) learn and innovate, (b) design and drive organizational simplicity, (c) manage ambiguity, and (d) manage and thrive on change.

Experimentation lies at the heart of every company's ability to innovate. What follows is the systematic testing of ideas, which is what enables

---

* The term *manager of innovation* (Moi) was coined by Frank Voehl and Jim Harrington in a workshop by the same name. The body of knowledge contains 10 competencies with which the Moi examines global trends in system innovation. He or she describes how to organize resources and flows, so that given targets, cost, clinical quality, and customer experience can be achieved with available resources. The Moi explores the importance of Lean process innovation, and presents new concepts, methods, and tools that can help to advance processes and operational flow. He or she also develops case studies of actual results in corporate innovation and involves concrete examples of the 10 types of innovation.

companies to create and refine their products. In fact, no product can be the product that it is without having first been an idea that was shaped, to one degree or another, through the process of experimentation. Today, a major development project can require literally thousands of experiments, all with the same objective: to learn whether the product concept or proposed technical solution holds promise for addressing a new need or problem, and then incorporating that information in the next round of tests so that the best product ultimately results.

Some initial CAI* ideas and concepts focused on assisting product designers in the early stage of the design process, but now a more comprehensive vision conceives CAI systems as beginning at the fuzzy front end of perceiving business opportunities and customer demands, then continuing during the creative stage in developing inventions. This is often followed by providing help up to the point of turning inventions into successful innovations in the marketplace.

CAI methods and tools are partially inspired by innovation theories, such as the theory of inventive problem solving (TRIZ), quality function development (QFD), axiomatic design, synectics, general theory of innovation, mind mapping, brainstorming, lateral thinking, and Kansei engineering, among others. The goal of these new CAI tools under development is to assist innovators, inventors, designers, process developers, and managers in their creative performance, with the expectation of changes in paradigms through the use of this new category of software tools. CAI, therefore, stands out as a departure from the usual trends.

As Paolo Gaudiano, president of Icosystem, noted, "ten years ago maybe one in twenty prospective clients had heard about agent-based simulation. Today it's about one in five. Ten years from now you'll be able to buy shrink-wrapped agent-based simulation software at Staples." That is why the future of innovation is simulation. Whereas before, we would sit among ourselves and decide how the world might work and test our ideas in the market, now we can test them in a virtual environment built by real-world data at much lower levels of cost and risk.

---

* Noel Leon, The future of computer-aided innovation, *Computers in Industry*, vol. 60, no. 8, pp. 539–550, 2009. Noel Leon is a full professor in the Centre for Innovation in Design & Technology at the Technological Institute of Monterrey, Mexico. His research work is oriented to CAI, product concept development, eco-innovation, computer-aided design and computer-aided engineering, and TRIZ (theory of inventive problem solving).

HOW TO USE THE TOOL

As such, an innovation simulation can be used to (a) define an opportunity or problem, (b) introduce the variables associated with the opportunity/ problem, (c) construct an innovation simulation model, (d) set up possible courses of action for testing, (e) run the simulation experiment, (f) consider the results, and (g) decide on a course of action. Innovation scenarios tackled by simulation can range from quite simple to complex, from grocery market checkout lines to analysis of the global economy. At its ideal, it helps us spot the innovation-related design problems earlier through rapid experimentation, as shown in Figure 17.1.

Adopting a simulation approach to address the rapid development of change management competencies related to product and process innovation has been revealed to have a number of important advantages, such as *solving problems earlier.* As Figure 17.1 shows, simulation *games* make it possible for managers to have a first-hand experience of how difficult it is to move from *theory to practice* or *from strategy to implementation.*

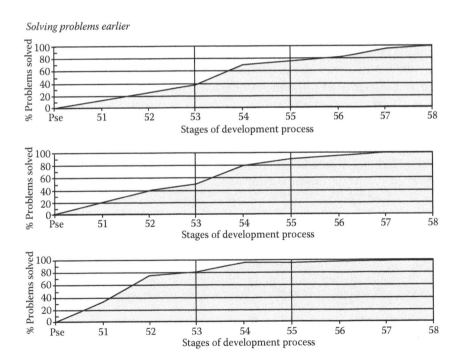

FIGURE 17.1
Impact of rapid experimentation.

As Figure 17.1 clearly shows, what can appear straightforward in the planning and development phases ends up being much more complex when we start interacting with actual (or simulated) people, as their reactions might not necessarily correspond to what we expected. Second, such simulations allow us to bring managers in touch with a wide variety of attitudes and behaviors (the *stereotypical* managers they have to convince) as well as to gain insights into diffusion processes (from linear to epidemic) operating across the formal but also informal networks of power and influence present in every organization. Third, such realistic management games provide a *playground* in which one can collaboratively experiment with the use of different tactics, generating rich discussions about their appropriateness and the most effective way of facing different forms of individual and organizational resistance.

Finally, a *game* approach is a nonthreatening, risk-free way of assessing one's strengths and weaknesses when confronted with high-pressure situations in which managing expectations, persistence, and team dynamics play a crucial role in determining the success or failure of innovation-related projects ultimately aimed at having a significant impact on people's habits, mind-sets, and cultures.

In his book *Diffusion of Innovations*,* Rogers defines the diffusion process as one "which is the spread of a new idea from its source of invention or creation to its ultimate users or adopters." Rogers differentiates the adoption process from the diffusion process in that the diffusion process occurs within society as a group process, whereas the adoption process pertains to an individual. Rogers defines the adoption process as "the mental process through which an individual passes from first hearing about an innovation to final adoption." Accordingly, at the beginning of the simulation, all the targeted adopters of the proposed change are still *unaware*, and it is the task of the player(s) to help them reach what Rogers

---

* In this renowned book, Everett M. Rogers, professor and chair of the Department of Communication and Journalism at the University of New Mexico, explains how new ideas spread via communication channels over time. Such innovations are initially perceived as uncertain and even risky. To overcome this uncertainty, most people seek out others like themselves who have already adopted the new idea. Thus, the diffusion process consists of a few individuals who first adopt an innovation, then spread the word among their circle of acquaintances—a process that typically takes months or years. But there are exceptions: use of the Internet in the 1990s, for example, may have spread more rapidly than any other innovation in the history of the world. Furthermore, the Internet is changing the very nature of diffusion by decreasing the importance of physical distance between people. *Diffusion of Innovation* addresses the spread of the Internet, and how it has transformed the way human beings communicate and adopt new ideas.

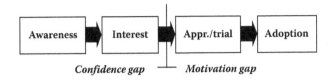

FIGURE 17.2
Knowing–doing gap.

calls the *awareness* stage. In the awareness stage, the individual is exposed to the innovation but lacks complete information about it.

Once awareness is reached, the *interest* stage of the simulation is entered. At the interest or information stage, the individual becomes interested in the new idea and seeks additional information about it. The next stage is *appraisal/trial,* a phase corresponding to Rogers' two stages of *evaluation* and *trial.* At the evaluation stage in the simulation, the individual mentally applies the innovation to his present and anticipated future situation, and then decides whether or not to try it. During the trial stage, the individual makes full use of the innovation. The last stage is *adoption.* At the adoption stage, the individual decides to continue the full use of the innovation. A particularly important transition is the one between *interest* and *appraisal/trial,* as here the targeted individuals need to be willing and able to start experimenting (and therefore take the risk of failing) with the new change. As discussed in Pfeffer and Sutton (2000), this transition can be easily linked to the above-mentioned work of Pfeffer and Sutton (2000) on the so-called knowing–doing gap (see Figure 17.2).

The simulation also tries to convey the importance of considering that the adoption process takes place differently in different people, as a function of (a) their unique specificities as individuals (history, personalities) and (b) their initial attitude toward change.

The preliminary results from the 10-year 1997–2007 Study on Complexity show that the number one source of complexity is the failure of organizations to integrate innovations and related changes into their work. It is the combination of these negative results and the cry that change will continue at an ever faster rate that has led to the proliferation of publications, consulting services, and training initiatives promoting good innovation management competencies and related simulation practices to practitioners in order to improve the success rate, as shown in Figure 17.3.*

---

* Ibid.

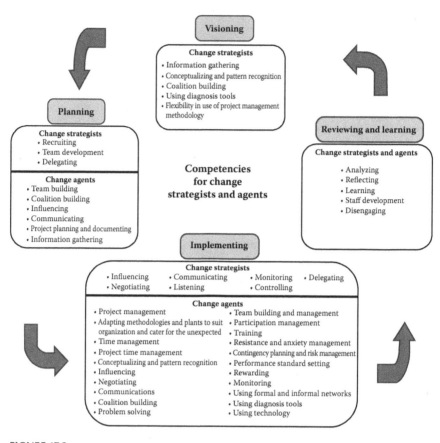

FIGURE 17.3
Competencies for innovation strategists and change agents.

During the planning stage of an innovation project, both the change strategists and change agents are active. The change strategists use simulations to improve their recruiting and team development skills, in order to build a cohesive project team and start delegating once the change project manager is recruited (Angehrn and Atherton, 1999). It is rare that one person has the breadth and depth of knowledge and competence necessary to envision the change or implement a major change. Therefore, the ability to use simulations shown in Figure 17.3 to recruit, build, and motivate teams of individuals with complementary strengths is crucial to success:

1. Understanding enlightened experimentation requires an appreciation of the process of innovation. We understand that product and technology innovations do not drop from the sky; they are nurtured

in laboratories and development organizations, passing through a *system* for experimentation.

2. All development organizations have such a system in place to help them narrow the number of ideas to pursue and then refine that group into what can become viable products. A critical stage of the process occurs when an idea or concept becomes a working artifact, or prototype, which can then be tested, discussed, shown to customers, and learned from.

3. The idea behind simulation is to try to imitate a real-world situation with a mathematical, physical, or virtual model that does not affect operations. The model should safely provide insights into outcomes of real-world scenarios at a minimal cost, at high levels of reliability when compared with other methods of analysis.

4. Innovation workshops or sessions based on the simulation start with a brief introduction before breaking the participants into teams and starting the simulation. The introduction typically highlights

   - Why being able to manage change is necessary in today's business environment and the apparent levels of corporate effectiveness in doing so
   - How individuals usually adopt change or innovation at differing rates and are sensitive to different approaches
   - How each virtual manager in the simulation has been modeled with backgrounds, personalities, and the ability to react to initiatives and tactics used by the change agent team as well as to other events—such as the influence of other managers in the virtual organization
   - What information and communication-based initiatives and tactics are available to them
   - How they will receive immediate feedback about the effects of their actions
   - The necessity for developing a strategy for how their change team will approach the implementation before *getting stuck into it*
   - A reminder that they could hear a buildup of frustration from some of the teams as they implement their strategy, for they are likely to encounter familiar hostile behavior and resistance patterns as found in the *real world*

5. Questioning techniques enable the participants to discuss and reflect on their own experiences as innovation change strategists, innovation change agents, or change recipients, including their own resistance to

change techniques. They are also challenged to think about how they will now handle situations upon return to their own organizations.

6. To take the learning experience from single-loop learning into the more effective double-loop learning process for each participant, debriefing sessions are used to link the learning generated from the simulation to the participants' personal experience of change and resistance. The following are four dimensions of the core capability for innovation, as depicted in Figure 17.4.

 a. *Employee knowledge and skill:* This dimension is the most obvious one.

 b. *Physical technical systems:* Technological competence accumulates not only in the heads of people; it also accumulates in the physical systems that they build over time—databases, machinery, and software programs.

 c. *Managerial systems:* The accumulation of employee knowledge is guided and monitored by the company's systems of education, rewards, and incentives. These managerial systems—particularly incentive structures—create the channels through which knowledge is accessed and flows; they also set up barriers to undesired knowledge-creation activities.

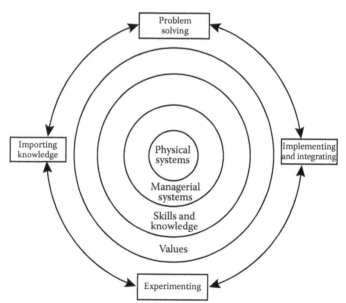

FIGURE 17.4
Dimensions of a core capability for innovation.

d. *Values and norms:* These determine what kinds of knowledge are sought and nurtured, what kinds of knowledge-building activities are tolerated and encouraged. There are systems of caste and status, rituals of behavior, and passionate beliefs associated with various kinds of technological knowledge that are as rigid and complex as those associated with religion. Therefore, values serve as knowledge-screening and -control mechanisms.

Figure 17.4 depicts the four dimensions of core capability, and relationships to systems, skills, knowledge, and value. Although at least potentially, some aspects of these four dimensions may be absorbed by outsiders, it is those portions of the system—and especially the synergy from unique combinations of them—that are neither readily transferable nor initiated. It is these dimensions that provide the organization with its strategic competitive advantage.

7. As reported in more detail in a recent qualitative study on the effectiveness of the information technology–based simulation and raised during the debriefing sessions when using the simulation, the learning points about managing change nearly always encompass

- Surprise at how realistically the simulation captures people and their reactions in a normal business situation
- Understanding of the need to identify and communicate with key stakeholders and key influencers
- Understanding the importance of information gathering
- Appreciation of the power of informal networks
- Awareness that receptiveness and resistance to change are different for each person
- Understanding that communication flows must be maintained in all directions
- Understanding the importance for maintaining momentum of change
- Understanding the importance and dynamics of teams
- Understanding the importance of having a strategy and then being flexible and able to adapt that strategy based on the feedback being received
- Appreciating the importance of and differences between time management and project time management

- Understanding the need to consider the impact of your strategy and tactics on all of the organization and on all stakeholders, not just the person you originally involve
- Appreciation of the effectiveness of a combined top–down, bottom–up strategy

Traditionally, the formal modeling of systems has been via a mathematical model, which attempts to find analytical solutions enabling the prediction of the behavior of the system from a set of defined parameters and initial conditions. Computer simulation is often used for modeling systems for which simple solutions are not possible. There are many different types of computer simulation. The common feature they all share is the attempt to generate a sample of representative scenarios for a model in which a complete determination of all possible states would be prohibitive or impossible.

How to Use the Strategy of Innovation

Organize for Rapid Experimentation

The ability to experiment quickly is integral to innovation: as developers conceive of a multitude of diverse ideas, experiments can provide the rapid feedback necessary to shape those ideas by reinforcing, modifying, or complementing existing knowledge. Rapid experimentation, however, often requires the complete revamping of entrenched routines. When, for example, certain classes of experiments become an order of magnitude cheaper or faster, organizational incentives may suddenly become misaligned, and the activities and routines that were once successful might become hindrances.*

Fail Early and Often, but Avoid Mistakes

Experimenting with many diverse—and sometimes seemingly absurd—ideas is crucial to innovation. When a novel concept fails in an experiment, the failure can expose important gaps in knowledge. Such experiments are particularly desirable when they are performed early on, so that unfavorable

---

* See the HBR paper "Enlightened experimentation: The new imperative for innovation," by Stefan Thomke (http://iic.wiki.fgv.br/file/view/Enlightened+Experimentation.pdf/221834486/Enlightened +Experimentation.pdf).

options can be eliminated quickly and people can refocus their efforts on more promising alternatives. Building the capacity for rapid experimentation in early development means rethinking the role of failure in organizations. Positive failure requires having a thick skin, says David Kelley, founder of IDEO, a leading design firm in Palo Alto, California.

IDEO encourages its designers "to fail often to succeed sooner," and the company understands that more radical experiments frequently lead to more spectacular failures.*

Anticipate and Exploit Early Information

When important projects fail late in the game, the consequences can be devastating. In the pharmaceutical industry, for example, more than 80% of drug candidates are discontinued during the clinical development phases, where more than half of total project expenses can be incurred. Yet, although companies are often forced to spend millions of dollars to correct problems in the later stages of product development, they generally underestimate the cost savings of early problem solving.

Simulation studies of software development, for instance, have shown that late-stage problems are more than 100 times more costly than early-stage ones, from a cost-of-innovation framework alone. For other environments that involve large capital investments in production equipment, the increase in cost can be orders of magnitude higher. In addition to financial costs, companies need to consider the value of time when those late-stage problems are on a project's critical path, as they often are. In pharmaceuticals, shaving six months off drug development means effectively extending patent protection when the product hits the market. Similarly, an electronics company might easily find that six months account for a quarter of a product's life cycle and a third of all profits.

Combine New and Traditional Technologies

New technologies that are used in the innovation process itself are designed to help solve problems as part of an experimentation *system*. A

---

* IDEO has developed numerous prototypes that have bordered on the ridiculous (and were later rejected), such as shoes with toy figurines on the shoelaces. At the same time, IDEO's approach has led to a host of bestsellers, such as the Palm V handheld computer, which has made the company the subject of intense media interest, including a *Nightline* segment with Ted Koppel and coverage in *Serious Play,* a book by Michael Schrage, a co-director of the event.

company must therefore understand how to use and manage new and traditional technologies together so that they complement each other. In fact, research by Marco Iansiti of Harvard Business School has found that, in many industries, the ability to integrate technologies is crucial to developing superior products.

A new technology often reaches the same general performance of its traditional counterpart much more quickly and at a lower cost. But the new technology usually performs at only 70% to 80% of the established technology. For example, a new chemical synthesis process might be able to obtain a purity level that is just three-quarters that of a mature technique. Thus, by combining new and established technologies, organizations can avoid the performance gap while also enjoying the benefits of cheaper and faster experimentation (see Figure 17.5).

Stephan Thomke, Harvard Business School, shows that new technology will reach perhaps just 70% to 80% of the performance of a traditional technology. A new computer model, he argues, might be able to represent real-world functionality that is just three-quarters that of an advanced prototype model. To avoid this performance gap, and potentially create new opportunities for innovation, companies can use the new and traditional technologies in concert. The optimal time for switching between the two occurs when the rates of improvement between the new

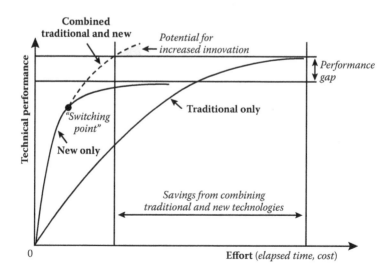

FIGURE 17.5
Model for technology combination for optimum innovation balance.

and mature technologies are about the same—that is, when the slopes of the two curves are equal.

Advantages of Simulation for Innovation

1. Simulation can be straightforward and flexible.
2. Software advances make simulation models easy to develop. Increased computer processing power at lower costs makes these simulation models more attractive.
3. Simulation can be used to analyze large and complex real-world situations that cannot be solved by conventional quantitative analysis models.
4. Simulation allows "what if?" types of questions.
5. Simulations do not interfere with the real-world system and ongoing operations.
6. Simulations allow us to study the interactive effect of individual components or variables to determine which ones are important.
7. Simulations allow for time compression.
8. Simulation allows for the inclusion of real-world complications that most quantitative analysis models cannot permit.

Disadvantages of Simulation for Innovation

- New simulation technologies can slash the costs (both financial and time) of experimentation and dramatically increase a company's ability to develop innovative products. To reap those benefits, though, organizations must prepare themselves for the full effects of such technologies, which is where simulation comes up short.
- Computer simulations and rapid prototyping, for example, increase not only a company's capacity to conduct experiments but also the wealth of information generated by those tests. That, however, can easily overload an organization that lacks the capability to process information from each round of experiments quickly enough to incorporate it into the next round.
- Cost may be prohibitive.
- The trial-and-error nature may not provide for optimal solutions.
- All relevant conditions and constraints must be generated for solutions to be examined.
- Each simulation model is unique, which can become a disadvantage in learning outcomes.

Another important goal of manufacturing-related simulations is to quantify system performance. Common measures of system performance include the following:

- Throughput under average and peak loads
- Cycle time
- Efficiency and utilization of resources, labor, and equipment
- Bottlenecks or constraints
- Queuing at work locations
- Queuing and delays caused by material-handling devices and systems
- Inventory storage requirements
- Transportation requirements
- Staffing requirements
- Effectiveness of scheduling systems
- Effectiveness of control systems

Advanced innovation tools like computer simulations can significantly increase innovators' problem-solving capacity, as well as their productivity, enabling them to address categories of problems that would otherwise be impossible to tackle. This is particularly true in the pharmaceutical, aerospace, semiconductor, and automotive industries, among others.* State-of-the-art simulation tools can enhance the communication and interaction among communities of innovators and practitioners, even those who are distributed in time and space. In short, new development tools (particularly those that exploit information technology) hold the promise of being faster, better, and cheaper—which is why companies like Intel and BMW have made substantial investments in innovation simulation technologies.

---

* For instance, computer simulation does not simply replace physical prototypes as a cost-saving measure; it introduces an entirely different way of experimenting that invites innovation. Just as the Internet offers enormous opportunities for innovation—far surpassing its use as a low-cost substitute for phone or catalog transactions—so does state-of-the-art experimentation. However, realizing that potential requires companies to adopt a different mind-set. Indeed, new technologies affect everything, from the development process itself, including the way a research and development organization is structured, to how new knowledge—and hence learning—is created. Thus, for companies to be more innovative, the challenges are managerial as well as technical.

SUMMARY

The new type of simulations for innovations can help us address efficiently other contexts related to people dynamics by enabling managers to experience difficult meeting situations in which virtual characters display different types of noncooperative behavior, or in which dysfunctional group dynamics, if not well handled, might threaten the outcome of important decision-making or negotiation sessions.

Ultimately, the models underlying these new innovation management development tools might also be embedded in current computer games played by children and young adults. In fact, role-playing and interactions with virtual characters in such games are still typically limited to shooting and fighting (a highly discussable approach to induce change) and might definitely profit from integrating equally challenging and entertaining, but more socially acceptable ways to have an impact on people's attitudes and behavior toward innovation, both online and offline.

The same concepts underlying the innovation simulation are currently used in the context of research projects sponsored by the European Commission aimed at designing *community games* helping the heterogeneous members of urban communities to better understand innovation dynamics in their own contexts. Involving the citizens of a town into simulation games of this type is expected to provide the stimulus for a broader reflection on how and why such communities adopt or reject innovation.[*]

EXAMPLES

Monte Carlo Simulation

When a system involves elements of chance in behavior, *Monte Carlo simulation* may be applied. The idea behind Monte Carlo simulation is to generate values that constitute the variables of the model under study. Very little in life is certain. There is a probabilistic nature to many of the actual systems that we might want to simulate (Law, 2014).

---

[*] A.A. Angehrn, Learning by playing: Bridging the knowing–doing gap in urban communities, in: A. Bounfour and L. Edvinsson, eds. *Intellectual Capital for Communities: Nations, Regions, Cities.* Burlington: Elsevier, 2005.

Some examples include (a) inventory demand over a specific time period, (b) lead time for inventory to arrive after order, (c) times between equipment failure, (d) times between arrivals at a maintenance facility, (e) service times, (f) times to complete project activities, and (g) number of employees absent from work for specific period.

Monte Carlo simulation is experimentation on the chance elements through application of random sampling. The five steps to Monte Carlo simulation are:

1. Establish the probability distributions for important input variables.
2. Build a cumulative probability distribution for each of the input variables.
3. Establish an interval of random numbers for each variable.
4. Generate random numbers.
5. Simulate a series of trials.

See Figure 17.6.

The smart vapor retarder: An innovation inspired by computer simulations

Water management is the new trend in civil engineering. Since it is difficult to ensure perfect vapor- and watertightness of building components, a limited moisture ingress is acceptable as long as the drying process is effective enough to avoid moisture damage. Recent computer models for the simulation of heat and moisture transport are valuable tools for the risk assessment of structures and their repair or retrofit. Unventilated, insulated assemblies with a vapor-resistant exterior layer can accumulate water because winter condensation and summer drying are not balanced. The balance can be reestablished if the vapor retarder is more permeable in summer than in winter. Parametric computer studies have defined the required properties of such a vapor retarder. Developed according to the computed specifications, the smart vapor retarder shows a seasonal variation in vapor permeability of a factor of ten. The secret of this behavior lies in the humidity-dependent vapor diffusion resistance of the film material.

Authors:	Kuenzel, H.M. [Fraunhofer Inst. for Building Physics, Holzkirchen (Germany). Div. of Hygrothermics]
Publication Date:	1998-12-31
OSTI Identifier:	687653
Report Number(s):	CONF-980650-- Journal ID: ISSN 0001-2505; TRN: IM9944%%318
Resource Type:	Conference
Resource Relation:	Conference: 1998 ASHRAE summer annual meeting, Toronto (Canada), 20 Jun 1998; Other Information: PBD: 1998; Related Information: Is Part Of ASHRAE transactions 1998: Technical and symposium papers. Volume 104, Part 2, PB: 1511 p.
Publisher:	American Society of Heating, Refrigerating and Air-Conditioning Engineers, Inc., Atlanta, GA (United States)
Country of Publication:	United States
Language:	English
Subject:	32 ENERGY CONSERVATION, CONSUMPTION, AND UTILIZATION; BUILDING MATERIALS; MOISTURE; MATHEMATICAL MODELS; RISK ASSESSMENT; DAMAGE; PERMEABILITY; MASS TRANSFER; HEAT TRANSFER; PROTECTIVE COATINGS; THERMAL INSULATION; ROOFS

FIGURE 17.6
Smart vapor retarder.

Computer-Aided Engineering

Computer-aided engineering (CAE) systems allow engineers to simulate the performance of designs electronically, providing the means for faster, lower-cost analysis. By incorporating engineering equations into the software, these systems can aid engineers in the selection of design parameters. Advanced systems may suggest alternative design concepts based on inputs provided by the design engineer. CAE systems provide rapid feedback on the potential performance of designs, improve initial design quality, and may provide a substitute to expensive physical model building and prototype construction.

Operational Gaming

Operational gaming refers to simulation involving two or more competing players. Examples include military and business strategy games. Participants match their management and decision-making skills in hypothetical conflicts.

Systems Simulation

Systems simulation is similar to business strategy gaming in that it allows users to test various management and decisions to evaluate their effect on the operating environment. These systems include corporate operations, national economies, hospitals, and governmental policies and systems.

Health Care and Patient Safety

Simulation in health care creates a safe learning environment that allows researchers and practitioners to test new clinical processes and enhance individual and team skills before encountering patients. Many simulation applications involve mannequins that present with symptoms and respond to the simulated treatment, similar to flight simulators used by pilots. Simulation allows health-care practitioners to acquire the skills and valuable experience they need safely, in a variety of clinical settings, without putting patients at risk (Agency for Healthcare Research and Quality).

Simulation in Entertainment

Simulation games represent or simulate an environment accurately. They represent the interactions between the playable characters and the environment realistically. Computer-generated imagery (CGI) is the application of the field of three-dimensional computer graphics to special effects. CGI is used for visual effects because they are high in quality, controllable, and can create effects that would not be feasible using any other technology either because of cost, resources, or safety.

Simulation in the Supply Chain

Simulation in the supply chain represents one of the most important applications of simulation innovation potential because of the connected-systems approach found in the extended enterprise experimental system.* This technique represents a valuable tool used by supply-chain managers when evaluating the innovation effect of capital investment in physical facilities such as factory plants, warehouses, and distribution centers. Simulation can be used to predict the performance of an existing or planned system, and to compare a series of alternative solutions for a particular innovation product or process design problem.

CASE STUDY

Strategic innovation simulation: Back Bay Battery (BBB) by Willy C. Shih and Clayton Christensen (see HBS at http://www.hbs.edu/faculty/Pages/item .aspx?num=37262).

This online simulation allows students to play the role of a business unit manager at Back Bay Battery Company, who faces the dilemma of balancing a portfolio of innovation strategies across products in the rechargeable battery space. Players have to manage research and development (R&D)

---

* The notion of the *extended enterprise experimental system* and its optimization had its roots in the logistics movement in the supply-chain arena. Edison's objective of achieving great innovation through rapid and frequent experimentation is especially pertinent today as the costs (both financial and time) of experimentation plunge. Yet many companies mistakenly view new technologies solely in terms of cost cutting, overlooking their vast potential for innovation. Worse, companies with that limited view get bogged down in the confusion that occurs when they try to incorporate new technologies. See *Macrologistics Management* (CRC Press, 1998), by Voehl and Stein for more details.

investment trade-offs between sustaining investment in the unit's existing battery business versus investing in a new, potentially disruptive battery technology. The student must also decide which market innovation opportunities to pursue, each of which offers the student varying levels of market intelligence and differing short- and long-term payoff prospects. Students manage the innovation project portfolios during eight simulated years. Throughout the simulation, the student is forced to address a number of challenges, including timing and level of innovation investment across both mature and new businesses, choices regarding innovation-related market opportunities and inherent product performance characteristics, requirements to meet constraining financial objectives and constant trade-offs between investment options, all in the context of uncertain market information.

BBB Case Study Note: The entire simulation can be played in 1.5 to 2.5 *seat hours.* A facilitator's guide contains an overview of simulation screens/ elements as well as a comprehensive teaching note. Hardware requirements: computer with minimum 1024 × 768 screen resolution, high-speed Internet connection (DSL/cable modem quality), Windows 2000, XP, or Vista/Macintosh operating systems, Internet Explorer 6+/Firefox 2.0+ web browser with Javascript and cookies enabled, Flash player 9+ browser plug-in (users with earlier versions of flash will be notified automatically and given the option to upgrade; this is a free browser plug-in), and Microsoft Excel (optional).

Case Study Learning Objectives: To Understand How

a. Best opportunities for new products are not visible early on.

b. New or unfamiliar applications can appear unattractive at first glance, but often represent best long-term opportunity.

c. Timing and the level of R&D spending are difficult to gauge, and must be accounted for in the simulation.

d. Assessing emerging market opportunities is difficult using standard approaches.

e. Balancing dual requirements for simultaneously investing in core business and innovation is challenging.

f. Constraining financial criteria and an organization's impatience for growth can make innovation difficult.

g. Case study subjects covered: competitive strategy, military/R&D, operations, product evolution, and process simulation.

SOFTWARE

Some commercial software available includes but is not limited to

Simulation software tools:

- Arena simulation, Rockwell Automation: www.arenasimulation.com
- ProModel: www.promodel.com
- SIMUL8: www.simul8.com
- ExtendSim: www.extendsim.com
- SolidWorks, Dassault Systemes: http://www.solidworks.com/

Excel add-ins:

- @Risk, Palisade Software: www.palisade.com
- Crystal Ball, Oracle: www.oracle.com
- Risk Simulator: www.risksimulation.com
- XLSim: www.xlsim.com

REFERENCES

Angehrn, A.A. and Atherton, J.E.M. *A Conceptual Framework for Assessing Development Programmes for Change Agents*. Fontainebleau, France: Centre for Advanced Learning Technologies, 1999.

Law, A.M. *Simulation Modeling and Analysis* (McGraw-Hill Series in Industrial Engineering and Management). New York: McGraw-Hill, 2014.

Pfeffer, J. and Sutton, R. *The Knowing-Doing Gap*. Boston: Harvard Business School Publishing, 2000.

SUGGESTED ADDITIONAL READING

Laguna, M. *Business Process Modeling, Simulation and Design*. Boca Raton. FL: Chapman and Hall/CRC, 2013.

Verschuuren, G.M. *Excel Simulations*. Uniontown, OH: Holy Macro! Books, 2013.

SIMULATION EPILOGUE ON CORE COMPETENCIES*

While it is difficult to change a company that is struggling, it is next to impossible to change a company that is showing all the outward signs

---

* From the introduction to the first chapter of *Wellsprings of Knowledge: Building and Sustaining the Sources of Innovation*, by Dorothy Leonard-Barton.

of success. Without the spur of a crisis or a period of great stress, most organizations—like most people—are incapable of changing the habits and attitudes of a lifetime.

John F. McDonnell
McDonnell Douglas Corporation

How would you like to move from a house after 112 years? ... We've got 112 years of closets and attics in this company. We want to flush them out, to start with a brand new house with empty closets, to begin the whole game again.

Jack Welch
Chairman and Chief Executive Officer (CEO), General Electric

We have to be willing to cannibalize what we're doing today in order to ensure our leadership in the future. It's counter to human nature but you have to kill your business while it is still working.

Lewis Platt
Chairman and CEO, Hewlett-Packard

Torawarenai sunao-na kokoro, which means *Mind that does not stick.*

Matsushita
Founder, Matsushita Electric Industrial

18

Six Thinking Hats

Frank Voehl

CONTENTS

DEFINITION

Six Thinking Hats is used to look at decisions from a number of important perspectives. This forces you to move outside your habitual thinking style, and helps you to get a more rounded view of a situation. The thinking is that if you look at a problem with the Six Thinking Hats technique, then you will solve it using any and all approaches. Your decisions and plans will mix ambition, skill in execution, public sensitivity, creativity, and good contingency planning.

USER

This tool can be used by individuals, but its best use is with a group of four to eight people. Cross-functional teams usually yield the best results from this activity.

OFTEN USED IN THE FOLLOWING PHASES OF THE INNOVATIVE PROCESS

The following are the six phases of the innovative cycle. An X after the phase name indicates that the tool/methodology is used during that specific phase.

- Creation phase X
- Value proposition phase X
- Resourcing phase X
- Documentation phase X
- Production phase X
- Sales/delivery phase X
- Performance analysis phase

TOOL ACTIVITY BY PHASE

The Six Thinking Hats is a tool that can be used in all the phases of the innovation cycle when problems occur or when new approaches are being developed. It is particularly useful during the creation phase and the production phase.

HOW TO USE THE TOOL

Six Thinking Hats is an important and powerful technique. It is used to look at decisions from a number of important perspectives. This forces you to move outside your habitual thinking style, and helps you to get a more rounded view of a situation. Many people think from a very rational, positive viewpoint, which is part of the reason that they are successful. Often, though, they may fail to look at a problem from an emotional, intuitive, creative, or negative viewpoint. This can mean that they underestimate resistance to plans, fail to make creative leaps, and do not make essential contingency plans. Similarly, pessimists may be excessively defensive, and more emotional people may fail to look at decisions calmly and rationally.

The thinking is that if you look at a problem with the Six Thinking Hats technique, then you will solve it using any and all approaches. Your decisions and plans will mix ambition, skill in execution, public sensitivity, creativity, and good contingency planning.

This tool was created by Edward de Bono in his book *Six Thinking Hats*. de Bono (1999) contends that they have developed many excellent thinking tools for argument and analysis. At the same time, information technology methods are constantly improving. But they have developed few tools to deal with our ordinary everyday thinking—the sort of thinking we do in conversations and meetings. In fact, our traditional thinking methods have not changed for centuries, de Bono contends. While these methods are powerful in dealing with a relatively stable world (where ideas and concepts tended to live longer than people), they are no longer adequate to deal with the rapidly changing world of today where new concepts and ideas like Six Thinking Hats are urgently needed.

Understanding the Six Thinking Hats

You can use Six Thinking Hats in meetings or on your own. In meetings, it has the benefit of blocking the confrontations that happen when people with different thinking styles discuss the same problem. Each *thinking hat* is a different style of thinking, as outlined in Figure 18.1.

White Hat

> With the use of this thinking hat, you focus on the data available. Look at the information you have, and see what you can learn from it. Look for gaps in your knowledge, and either try to fill them or take account of them. In other words, this is where you analyze past trends, and try to extrapolate from historical data.

Red Hat

> Wearing the red hat, you look at problems using intuition, gut reaction, and emotion. Also try to think how other people will react emotionally. Try to understand the responses of people who do not fully know your reasoning.

Colored hat	Think of	Detailed description
	White paper	The white hat is about data and information. It is used to record information that is currently available and to identify further information that may be needed.
	Fire and warmth	The red hat is associated with feelings, intuition, and emotion. The red hat allows people to put forward feelings without justification or prejudice.
	Sunshine	The yellow hat is for a positive view of things. It looks for benefits in a situation. This hat encourages a positive view even in people who are always critical.
	A stern judge	The black hat relates to caution. It is used for critical judgement. Sometimes it is easy to overuse the black hat.
	Vegetation and rich growth	The green hat is for creative thinking and generating new ideas. This is your creative thinking cap.
	The sky and overview	The blue hat is about process control. It is used for thinking about thinking. The blue hat asks for summaries, conclusions and decisions.

FIGURE 18.1
Overview of the Six Thinking Hats.

Black Hat

Using black hat thinking, look at all the bad points of the decision. Look at it cautiously and defensively. Try to see why it might not work. This is important because it highlights the weak points in a plan. It allows you to eliminate them, alter them, or prepare contingency plans to counter them. Black hat thinking helps make your plans *tougher* and more resilient. It can also help you spot fatal flaws and risks before you embark on a course of action. Black hat thinking is one of the real benefits of this technique, as many successful people get so used to thinking positively that often they cannot see problems in advance. This leaves them underprepared for difficulties.

Yellow Hat

Using the yellow hat helps you to think positively. It is the optimistic viewpoint that helps you see all the benefits of the decision and the value in it. Yellow hat thinking helps you keep going when everything looks gloomy and difficult.

Green Hat

The green hat stands for creativity. This is where you can develop creative solutions to a problem. It is a freewheeling way of thinking, in which there is little criticism of ideas.

Blue Hat

The blue hat stands for process control. This is the hat worn by people who are team leaders and who are chairing meetings. When running into difficulties because ideas are running dry, they may direct activity into green hat thinking. When contingency plans are needed, they will ask for black hat thinking, etc. (see Figure 18.2).

Using the Six Thinking Hats

The directors of a property company are looking at whether they should construct a new office building. The economy is doing well, and the amount of vacant office space is reducing sharply. As part of their decision, they decide to use the Six Thinking Hats technique during a planning meeting.

1. Looking at the problem with the white hat, they analyze the data they have. They examine the trend in vacant office space, which shows a sharp reduction. They anticipate that by the time the office block would be completed, there will be a severe shortage of office space.

Questions to be asked

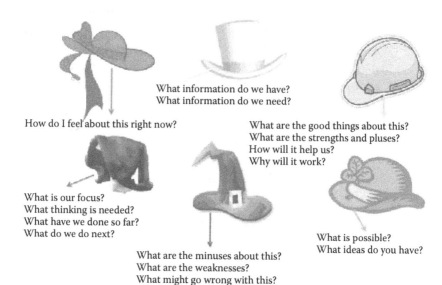

What information do we have?
What information do we need?

How do I feel about this right now?

What are the good things about this?
What are the strengths and pluses?
How will it help us?
Why will it work?

What is our focus?
What thinking is needed?
What have we done so far?
What do we do next?

What is possible?
What ideas do you have?

What are the minuses about this?
What are the weaknesses?
What might go wrong with this?

FIGURE 18.2
More insight into the Six Thinking Hats.

Current government projections show steady economic growth for at least the construction period.

2. With red hat thinking, some of the directors think the proposed building looks quite ugly. While it would be highly cost-effective, they worry that people would not like to work in it.

3. When they think with the black hat, they worry that government projections may be wrong. The economy may be about to enter a *cyclical downturn*, in which case the office building may be empty for a long time. If the building is not attractive, then companies will choose to work in another better-looking building at the same rent.

4. With the yellow hat, however, if the economy holds up and their projections are correct, the company stands to make a great deal of money. If they are lucky, maybe they could sell the building before the next downturn, or rent to tenants on long-term leases that will last through any recession.

5. With green hat thinking, they consider whether they should change the design to make the building more pleasant. Perhaps they could build prestige offices that people would want to rent in any economic

climate. Alternatively, maybe they should invest the money in the short term to buy property at a low cost when a recession comes.

6. The blue hat has been used by the meeting's chair to move among the different thinking styles. He or she may have needed to keep other members of the team from switching styles, or from criticizing other peoples' points.

See more at http://www.mindtools.com/pages/article/newTED_07.htm; also see http://www.storyboardthat.com/articles/business/brain-storming /six-thinking-hats#sthash.jfmC2lUo.dpuf.

Advantages

The Six Thinking Hats method has a lot of different advantages, which include the following:

- It leads to shorter and more productive thinking processes.
- It reduces conflict within a group and motivates team members to think more clearly.
- Learners learn how to look at problems and decisions systematically, how to achieve best results within a group situation, and how to improve the quantity as well as the quality of their ideas.
- Learners experience first hand how to solve problems with innovative and creative ideas, as well as how to spot potential threats to the situations and how to overcome or avoid them.
- Learners are taught how to see beyond the obvious ideas and to see all the sides of a given situation.
- de Bono's thinking hats method enhances collaborative thinking between learners in a group and reduces conflict by acknowledging everyone in the group's viewpoints.
- Six Thinking Hats is a simple yet powerful tool based on a principle of parallel thinking—everyone thinking in the same direction, from the same perspective, at the same time.
- It helps people step outside the confines of fixed positions and one way of thinking.
- Western thinking style is based on adversarial debate *with* people thinking and interacting from differing perspectives and positions.
- This tool enables us to look at things in a collaborative way.
- It also helps us see beyond our normal perspective to *visualize* new opportunities.

- To achieve the service improvements, you will have to change the way you do things. This means thinking up and considering new ideas.
- If you evaluate the change from a number of perspectives, you and your team will have a more rounded outlook on the ideas.

Disadvantages

- Probably the biggest disadvantage of de Bono's thinking hats is that the process of thinking can be very time consuming.
- Usually, when solving a problem or thinking about a situation, one thinks overall about it and the process goes very fast as one quickly runs out of ideas.
- When thinking from six different perspectives instead of one, it can take up a lot of time, which would have otherwise been available for additional tasks.
- Another disadvantage is that this method of teaching and thinking can have a negative effect on the learners that cannot work well in group situations. Often a lot of learners tend to think and work better on their own than in group situations, and learners like that will not benefit from this method.
- Another disadvantage is that there can still arise conflict between the learners' views of the different perspectives.
- A disadvantage can often be how you apply it into a real group meeting. At least you need time to show your group member how to do it and explain to them, then need some practice.
- For the long term, it is good; however, if you want to create a new idea or solve a problem for your group meeting in a short term, it may not work that well.

SUMMARY

The author views Six Thinking Hats as a good technique for looking at the effects of a decision from a number of different points of view. It allows necessary emotion and skepticism to be brought into what would otherwise be purely rational decisions. It opens up the opportunity for creativity within the decision-making process.

Six Thinking Hats

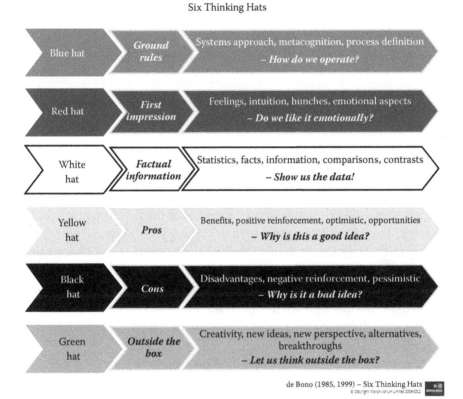

de Bono (1985, 1999) – Six Thinking Hats

FIGURE 18.3
Suggestions on using the Six Thinking Hats.

The technique also helps, for example, persistently pessimistic people to be positive and creative. Also, plans developed using the Six Thinking Hats technique will oftentimes be sounder and much more resilient than would otherwise be the case. It may also help you to avoid PR disasters and mistakes, and identify reasons not to follow a course of action before you have committed to it.

A summary of the Edward de Bono *Six Thinking Hats* is provided in Figure 18.3.

EXAMPLE

Examples are included in the section entitled "How to Use the Tool."

CASE STUDY

Emerson (Copeland Corp) Scroll Compressors Plant

Plant Managers Use Innovative Communication Tools to Tap Strengths of Straight-Shooting Workforce

Challenge: Growth Creates Communications Roadblocks

The Emerson Copeland facility in Ava, Missouri, had once been an Emerson motor plant. When that operation was moved to Mexico, Copeland bought the plant and brought some of the Emerson people back to work. The rehires were committed to keeping their jobs, and over time, owing to their performance, the plant expanded four times. The growth caused the division to restructure and the local plant name changed to Scroll Compressors LLC. Growth is usually a positive, but it brings new challenges. The quick growth the plant was experiencing had limited many employees' willingness to engage management. Employees, described by managers as a tremendous resource, honest and smart, as well as hard working, were not giving their opinions anymore.

Solution: Teach and Apply Communication and Problem-Solving Skills to Rekindle Honest Interaction between Managers and Employees

In a manufacturing plant, employee involvement requires commitment. Holding meetings takes employees off the floor, temporarily interrupting production. And listening to what hundreds of workers are experiencing and thinking takes a lot of management time. But Scroll Compressors LLC managers Michael Redfearn, Skip Steward, and Bill Henry believed that if you make the effort, you can harvest valuable information on problems that exist along with practical suggestions for solving them.

Working with the plant trainer, Don Hanger, and consulting firms including Solution Technologies, Strategy Associates, and de Bono Consulting, this management team designed a five-step plan to increase employee involvement in improving plant performance:

1. Learn de Bono communication and problem-solving tools and pilot the process.
2. Train senior management as facilitators.

3. Create Ava Impact Teams to address specific tasks/problems.
4. Gradually engage all of staff.
5. Measure and publish results.

The long-term goal was to enable Copeland to continue to provide superior-quality products at a reasonable price. The first short-term goal was to deal with scrap issues in a new automated-production area.

Value: Measurable Results Lead to Wider Application of the de Bono Tools

Using the de Bono Six Thinking Hats tool leveled the playing field between supervisors, engineers, and frontline workers. Communication improved, and positive energy began to flow. The ideas generated and implemented by impact teams **reduced scrap by 20%, resulting in an annual $48,000 cost savings**. As a result of this success, all of management was then trained in the methodology, and de Bono systems are now part of regular protocol.

An example of a subsequent impact team focus was a plant-wide contest to reduce electricity usage. One team examined the plant's practice of running coolant pumps on weekends to keep a water-based coolant moving, even though the machines it cooled were not in operation. The reason for this practice was to control bacterial growth, which otherwise produced a very foul odor. Applying the tools they had learned, the team came up with the idea of circulating the coolant in bursts, running the equipment only once for 5 to 10 minutes every three to four hours. Implementation of the burst approach **saved the plant $130,000 in one year**. The total **savings through all ideas implemented from the contest was $400,000**. At the time this case study was written, impact teams are focused on ways that the plant can become greener by keeping waste out of landfills.

Future: Involvement of a Workforce Rooted in a Small Western Town Leads to Growing Impact

Don Hanger reports that the Ava plant's employees use the de Bono tools for idea generation whenever they encounter something that they do not know how to deal with. The tools are also embedded in various Lean manufacturing processes such as Six Sigma and Total Preventive Maintenance to strengthen outcomes. The Six Thinking Hats methodology shown in Figure 18.4 is used routinely in performance reviews of hourly workers. In

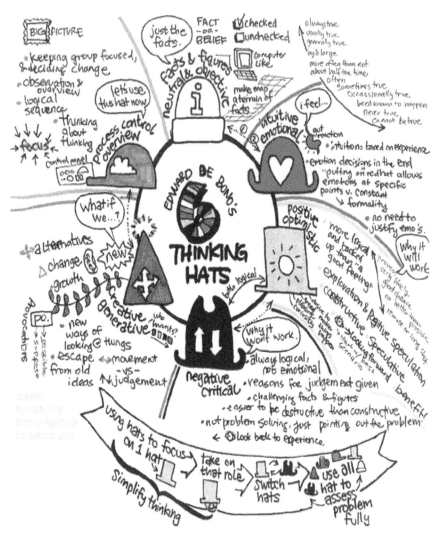

FIGURE 18.4
Emerson's view of the Six Thinking Hats.

fact, so many things are going on that it is becoming difficult to track all of the results that can be attributed to the use of this practical and innovative problem-solving system.

"In the future what I'd most like to see," Don concludes, "is for Emerson and its partners to make use of our demonstrated success by adopting the de Bono techniques in other manufacturing plants—and other divisions of the organization. It would be great to see an approach that was pioneered in the Ava plant benefit Emerson on a much larger scale." (See Figure 18.4.)

SOFTWARE

- *Tuzzit*—This Six Thinking Hats software is a creative thinking technique used to look at decisions from a number of different points of view. It forces the participants to move outside of their habitual thinking style, and helps in the decision-making process (see https://www.tuzzit.com/en/canvas/six_thinking_hats?gclid=CI3zjYzWwL8CFSqZMgodZ1cA-Q).
- *MindJet*—The platform for enterprise innovation. More than collaborative, more than social, Mindjet is the engine for innovation programs that keep the crowd engaged (see http://www.mindjet.com/).
- *Insight Maker*—Insight Maker runs in your web browser. No software download or plug-ins are needed. Get started building your rich pictures, simulation models, and Insights (for further information, see http://insightmaker.com/).
- *X-Mind*—You can use simple mind maps if you choose, or *fishbone*-style flowcharts if you prefer. You can even add images and icons to differentiate parts of a project or specific ideas, add links and multimedia to each item, and more (see http://www.xmind.net/).
- The Six Thinking Hats App on Google—This application is a quick way to remind you how to apply the six hats (http://www.parade.vic.edu.au/md/teacher_research_guide/defining_debono.htm).

Other commercial software available includes but is not limited to

- Storyboardthat.com
- www.edrawsoft.com
- www.smartdraw.com
- mobile.brothersoft.com
- www.qimacros.com

REFERENCE

de Bono, E. *Six Thinking Hats*. New York: Back Bay Books, 1999.

SUGGESTED ADDITIONAL READING

Asian Development Bank. *Learning for Change in ADB*. Mandaluyong City, Philippines: Asian Development Bank, 2009, p. 21. Available at http://www.adb.org/sites/default/files/publication/29000/learning-change.pdf.

de Bono, E. *Lateral Thinking: A Textbook of Creativity*. New York: Penguin Adult, 2001. ISBN 0141033088.

Harrington, H.J. and Lomax, K. *Performance Improvement Methods*. New York: McGraw-Hill, 2000.

Mattiske, C. *Creative Business Thinking E-book*, Amazon e-book.

19

Social Networks

Peter Westbrook and Neil Farmer

CONTENTS

This chapter has been produced in association with Paul Goodstadt ACIB, FRSA, FTQMC, director of development at the Total Quality Management College, Manchester, United Kingdom.

DEFINITION

Social networks are networks of friends, colleagues, and other personal contacts: strong social networks can encourage healthy behaviors. They are often computers or online communities of people with a common interest who use a website or other technologies to communicate with each other and share information, resources, etc. A business-oriented social network is a website or online service that facilitates this communication.

USER

This tool can be used by networks as small as two people or by extremely large networks of people.

OFTEN USED IN THE FOLLOWING PHASES OF THE INNOVATIVE PROCESS

The following are the seven phases of the innovative cycle. An X after the phase name indicates that the tool/methodology is used during that specific phase.

- Creation phase X
- Value proposition phase X
- Resourcing phase X
- Documentation phase X
- Production phase X
- Sales/delivery phase X
- Performance analysis phase X

TOOL ACTIVITY BY PHASE

Social networks can be used in all phases of the innovation process, as they are a communication tool as well as a problem-solving tool.

HOW TO USE THE TOOL

Social networks can change leadership and innovation within an organization. It is a lot like mending a broken DNA (see Figure 19.1).

This chapter is particularly important for business organizations and outsource providers as they approach major business changes.

More than 65,000 leadership books have been published containing many hundreds of ideas on how to become an effective leader. Despite all the good, bad, and indifferent leadership research that has been carried out, leadership (and particularly change leadership) today is not fit for purpose in most organizations.

For decades, some 70% of major change initiatives have failed to meet their key objectives; less than 30% of managers are influential with their subordinates and other key colleagues; and only about 15% to 20% of employees are really engaged with their organizations at any point in time, even less during periods of major change.

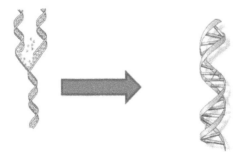

FIGURE 19.1
Mending a broken DNA.

What Is Effective Change Leadership? Sorting the Wheat from the Chaff

Before moving on to the question of how to solve the huge problems of major change implementation, the vast array of leadership and change ideas that whirl around leadership theory need to be cut down to size. Let us get practical: problems of the scale described above will not be solved by a bucket load of worthy but marginal improvements. We need to focus on a relatively few key insights to sort out the *wheat from the chaff* in practical, effective leadership.

- Big leadership insights

  The huge leadership problems mentioned above can only begin to be addressed once some fundamental truths are recognized and accepted:

  – There is a huge mismatch between the formally recognized, so-called leaders in management hierarchies and the *real* leaders in organizations. The real leaders come in various forms: those who strongly influence the views and behaviors of colleagues, those who others go to for knowledge and advice, those who collaborate effectively across organizational boundaries, those who are naturally innovative, and so on. Most of these real leaders are unrecognized, hugely underutilized, and not in management positions.

  – Influencers will *emerge* in all human groups who interact together. However, these key individuals are associated with their specific groups: move a strong innovative influencer from one group to another, and the continuation of influence is a lottery.

  – Less than 20% of managers are strong, effective, change-positive influencers. Although, on average, up to 30% of managers have some innovative influence credentials (in terms of influencing some colleagues and being at least open-minded on change), a more rigorous assessment (strongly influential and change-positive by nature) brings this figure down to less than 20%.

  – Transferring the skills and behaviors of successful (normally senior level) leaders to others through *leadership* training and development has a very mixed track record. The fundamental flaw is that (mostly) the wrong people are being developed. In addition, effective leadership development tends to follow the

70:20:10 rule: only about 10% of effective leadership skills and behaviors are learned through attending leadership courses, while 20% are accumulated through solving work-related problems/ exploring opportunities via personal work networks, and the bulk (70%) are absorbed by being stretched in projects and other activities outside of normal work boundaries.

- The most effective leadership of people occurs where innovative influencers are also in relevant formal leadership positions—where managers are leaders. When this happens, both performance and employee morale rise, even during periods of profound change. This is the single most effective way to consistently achieve and maintain *real* employee engagement.
- The most consistently effective innovation occurs when individuals with different skills and knowledge work together across relevant organizational boundaries, both internally and externally. While formal mechanisms are frequently necessary to provide a clear focus (such as a new product or service requirement or a specific problem to be solved), the key ingredient is almost always activating relevant interpersonal (largely informal) networks.
- In today's real world of work, consistently effective leadership is *networked*: effective managers will increasingly be those who work with both formal and informal leaders to achieve desired goals.

What These Insights Tell Us about Today's Change Programs

These insights enable us to get a clearer picture of what is really going on beneath the logic of program and project management methods in major change initiatives today. In essence, these leadership insights tell us that there are two strands to any change program: a formal strand and an informal strand. The formal strand contains all the logical, easily understood elements, such as project management methods, the technical skills needed to create new systems and processes, the people with the necessary competencies, more effective organizational designs, etc.

The informal strand contains all the relevant informal relationships that exist within the organization and beyond: those who influence colleagues, those who are experts and share their knowledge, those who are innovative, managers who are also influential, those who act as communication bridges between different groups, and so on.

Today's change programs largely ignore the informal strand, leading to a huge variety of people problems, including

- Noninfluential individuals are chosen by noninfluential managers to act as change champions and user representatives on change teams, leading to low levels of buy-in by colleagues.
- Change communications is largely top–down, leading to widespread cynicism, a lack of cooperation, and communication gaps being left to fester and grow.
- A scramble for some future roles (often with focused *backstabbing*).
- Restructuring often leads to the breaking up of highly effective informal networks—teams where the manager is very influential with team members, cross-area innovation networks, knowledge transfer networks based on key individuals, and a variety of established interteam collaboration networks.
- Fearful staff drift toward less risky roles.
- Lack of flexibility between teams.
- More and more cost and timescale overruns.
- The most effective individuals become disillusioned, dust down their curriculum vitaes, and leave (a trickle at first, then in droves).
- Staff are afraid of pointing out genuine problems with new ways of working—"just doing what we are told."
- Change program objectives are progressively scaled down to match the disappointing realities.

What These Insights Tell Us about Effective Change Programs

Rather like the double-helix structure in DNA, effective change programs synchronize formal and informal strands. Formal elements, such as program and project management methods, technical and restructuring specialists, etc., are integrated with additional informal elements that are missing from almost all change programs today. While people problems do not disappear in this integrated formal/informal approach, they are seriously reduced:

- Change-positive and open-minded influencers are chosen through peer surveys to act as change champions and user representatives on change teams, leading to high levels of buy-in by colleagues.

- Change communications is a synchronized combination of bottom-up and top–down channels, with heavy involvement of key influencers, an open and honest environment, and extensive feedback. This approach leads to widespread confidence that the truth is being told and questions answered honestly, leading to routine cooperation, with communication gaps being recognized early and remedied through both formal and informal channels.
- An open process for selection of people for future roles, based on both their formal and informal (network) strengths: generally without any backstabbing.
- A *do no harm to key informal networks* approach to restructuring. For example, this means that change-positive and open-minded influencers should be retained. This is particularly true where these key individuals strongly influence the teams they work with: in these situations, both the influencers and their teams should be kept intact during any restructuring. This principle also applies to key knowledge sharing and innovation networks. Typically, this means that some 10% to 20% of staff in organizations will remain unchanged as groups at the end of the restructuring—they may report into different managers or areas but the groups remain intact. (Implementers of mergers, take note.)
- Key staff is willing to take on more challenging roles.
- Increased flexibility between teams.
- Very high levels of outcomes within timescales and budgets.
- The most effective individuals become empowered and few leave the organization.
- Staff is not afraid to point out genuine problems with new ways of working.
- Change program objectives are progressively achieved, matching the *above expectation* realities.
- The principles listed in Figure 19.2 are essential to success.

The small number of major change programs that have used this synchronized approach have an exceptionally good track record, with more than 90% of the *original* change objectives being achieved, almost always on time and on budget. Two very different examples are summarized in the succeeding pages ("National Savings: 100% Productivity Improvement Using Key Influencers" and "Friends Provident: Profound Culture Change in Less than a Year").

Success on this scale can only be achieved through the discretionary effort of key influencers and other informal leaders. The following principles are key to success:

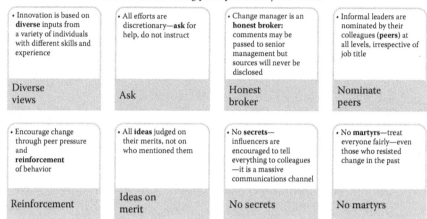

FIGURE 19.2
Principles that are key to success.

National Savings: 100% Productivity Improvement Using Key Influencers

The single biggest private finance initiative (PFI) contract issued by the UK government at the time involved a practical partnership between National Savings (investment assets: £60 billion) and Siemens Business Services (SBS), part of the giant Siemens group. Under this contract, SBS runs all the administrative systems and processes for National Savings, as well as takes over responsibility for some 4000 staff. The objective of this PFI was nothing less than to create a highly effective, automated, and modern organization that can fully meet the current and future needs of National Savings, while at the same time utilizing surplus staff and a comprehensive information technology/information systems (IT/IS) infrastructure to service new business clients. The technology involved includes e-commerce, call centers, and intelligent document scanning and recognition.

SBS has significantly increased productivity (more than 100%) without compulsory redundancies through reducing the number of staff transferred to it from 4100 to some 2000, although the volume of work has remained the same.

National Audit Office, 2004
Victoria, London

Friends Provident: Profound Culture Change in Less than a Year

The IT department at Friends Provident embarked on a radical change initiative to transform a relatively traditional IT department into an internal *commercial IT service* that really treated the Friends Provident business as its customer. This change required organizational, process, and internal charging changes that were implemented efficiently during a period of six months using traditional methods. However, it rapidly became evident that a radical change was needed to the culture of the IT group—relationships with customers, the way that new processes were implemented in practice, and (above all) the way that IT staff behaved in their day-to-day work. We led a nine-month program of practical culture change, using *waves* of managers and influencers to identify and then progressively *live* and reinforce new desired behaviors.

The end result was a profound and measurable change in "the way that we do things here," with staff consistently carrying out the desired behaviors increasing from 18% to 55% during the first nine months, with another 30% following suit in the next three months. These culture changes were reflected in significantly improved customer satisfaction and staff satisfaction ratings.

The Gartner Group wrote a case study of the successful Friends Provident experiences and the director of IT prepared a video describing their experiences. A year later, in 2007, the Friends Provident IT Group won the *Computer Weekly* "Best place to work in IT" award for IT in financial services.

Why Don't Most Major Change Programs Adopt This Synchronized Approach?

As mentioned above, there is a huge mismatch between the formally recognized, so-called leaders in management hierarchies and the *real* leaders in organizations.

Definition: A *real leader* is an individual who is both highly influential with relevant colleagues and capable of sound judgment in making decisions/forming views relevant to their role and to the broader business environment.

Our experience of using the formal/informal synchronized approach in a variety of organizations employing more than 26,000 staff in total clearly shows that the most influential people are found within the largest work groups: some key influencers in large work groups influence the views of up to 50 people. Managers (particularly senior managers who generally work day-to-day with fewer colleagues) are much less likely to influence large numbers of people in the organization. A typical manager who is also a *leader* is likely to influence some 10 to 15 colleagues while having the capability to make sound decisions relevant to their role.

Now consider a typical senior manager in a large organization today: he or she has no experience of the formal/informal synchronized approach. Senior managers are therefore very reluctant to accept influence realities and will only slowly adopt the most effective change management approach available. The sooner they learn about and start adopting synchronized change implementation, the sooner they (or at least some of them) will become real leaders.

Why Don't Program/Project Management Consultants Use the Synchronized Approach?

Most large management consultancies are familiar with the theory, and sometimes even the limited use of informal networks for change. However, they are constrained by two main factors:

1. A lack of understanding and demand for the approach from clients.
2. The fact that the synchronized approach generally requires considerably fewer consultants to implement change initiatives. Typically, a single consultant can coordinate the activities of some 50 key influencers, who in turn influence some 20 different colleagues on average. Thus, there is no need for coachloads of change management consultants (see Figures 19.3 and 19.4).

Will Outsource Providers Adopt the Synchronized Approach to Change?

Most of our successful change implementations using the synchronized formal/informal approach for major change programs have been for

Informal networks survey analysis: Typical change influencer *nomination* patterns

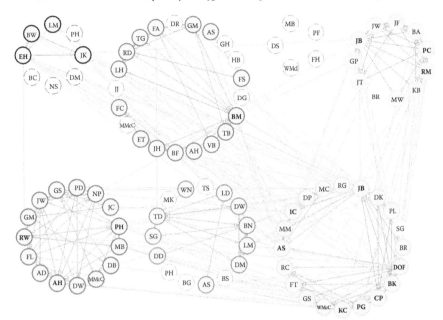

Note: Individual initials are shown in the small circles with different business areas being color coded

FIGURE 19.3
Information survey analysis.

Gearing up change through key
influencers: The 1:1000 effect

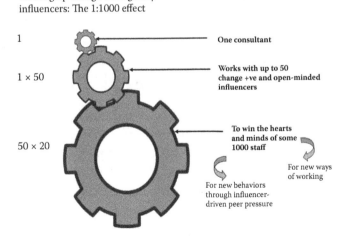

FIGURE 19.4
Key influencers.

outsource providers. When an outsource provider takes over a large number of staff as part of an outsource contract, many of the senior management perceptions about the approach disappear. The provider has no illusions about who the real leaders are in the transferred organization: unlike many client senior managers, they know what they do not know. In these situations, there are many attractions in identifying the real *movers and shakers* at an early stage. Bringing in a specialist to implement change in subsequent change programs is then a natural next step.

Both in terms of offering a competitive product and subsequent bottom-line outcomes, we believe that the synchronized approach is truly strategic for outsource suppliers. It can (with confidence) boldly go where client senior managers today generally fear to tread.

- The transfer of significant numbers of client staff into a totally new organization under TUPE arrangements (no changes to employee terms and conditions for two years) is an opportunity to create a smaller but more motivating work environment—by significantly increasing the number of managers who are also leaders.
- The commercial necessity to reduce costs/improve efficiency, etc., as quickly as possible can be achieved through involving influencers and other informal leaders to win *hearts and minds* through open-and-honest communication from day one.
- Building good working relationships with clients, including those staff who are responsible for managing the outsource delivery key performance indicators, etc., is a natural by-product of effective, open, and honest change management—feedback from former colleagues is a major factor in building trust into the client/provider relationship.
- Implementing new technology to enhance systems and processes. The challenge here is often in motivating and reengaging with the staff that remain after these profound implementations. The synchronized formal/informal approach is a very powerful and effective solution to all aspects of staff engagement going forward.

There have been many false dawns for the use of informal networks in organizational change. Despite this, we believe that the use of the synchronized formal/informal approach is so powerful and effective that its time is ripe: by 2020, this approach will be in widespread use. In the meantime, outsource providers will often be in the vanguard.

Using the Synchronized Approach
for Consistently Effective Innovation

As mentioned above, the most consistently effective innovation occurs when individuals with different skills and knowledge work together across relevant organizational boundaries, both internally and externally. While formal mechanisms are frequently necessary to provide a clear focus (such as a new product or service requirement or a specific problem to be solved), the key ingredient is almost always activating relevant interpersonal (largely informal) networks.

- Collaboration across formal and informal organizational networks

  From an organizational perspective, some areas in the organization will work most heavily through informal personal networks, particularly innovation projects, business development activities, and related change implementations. However, all areas of the organization will function to a significant degree through informal networks, particularly during innovation refinement and implementation.

  In any serious innovation strategy, care needs to be taken to ensure that the resources (people and money) allocated to all types of innovation—continuous improvement, new products/services, and real breakthrough/radical innovations—are in balance, allowing all to thrive.

  At a strategic level, a small number of external specialists (with scientific, technical, market awareness, and other related skills) will work on projects that also involve key individuals with internal technical and sales/marketing skills. In addition to these *obvious* choices, however, a small number of internal influencers and innovators from operational areas should be used to seamlessly enhance the entire innovation/implementation process.

  Strategic innovation projects often spawn a series of follow-on projects that refine the strategic options into practical actions and implementation sequences. During these follow-on projects, the external specialists will normally be absent, with internal *experts*, innovators, and key influencers playing major roles.

- Choosing the right people

    The first step in implementing a practical, integrated formal/informal design for innovation in an organization is to choose the right people, with very different but relevant backgrounds and skills. Two main guidelines should be employed:

    - Formal innovation managers should be chosen primarily for a natural collaborative style, with training and experience in innovation being a secondary consideration.
    - Individuals in all innovation roles should be chosen not just on their innovative skills and relevant knowledge, but also on the quality of their personal networks. In this way, you will rapidly be able to assemble not just capable individuals but a whole set of collaborative networks to deliver effective accelerated innovation. This is hugely important to the future success of innovation in any organization.

Engaging Emergent Leaders within a *Bounded Freedom* Environment

The most effective approach to innovation is to set clear boundaries: a target product or service set, resources, and timescales. Within these boundaries, however, the innovation teams will be largely self-organizing.

Teams start with a transitory project manager, with knowledge of/access to all relevant research and outcomes from previous innovation projects. However, this individual's mandate is simply to set the boundaries and act as project initiator/manager only until a set of natural leaders emerge over the first 4–6 weeks of each project. This emergent process requires considerable face-to-face interactions in the early stages, with virtual interactions becoming more common thereafter. Experience indicates that after an initial multimeeting period of face-to-face interactions (weeks in total not days), at least one in four of follow-up meetings should also be face-to-face rather than virtual.

The most influential emergent leader will usually (but not always) take over the project manager role, with other natural leaders being given roles relevant to their natural abilities: lead innovator, lead researcher, lead subject expert, etc. In exceptional situations, where no clear natural leader emerges within the team, the transitory project manager may stay in place as a pragmatic solution for that team (see Figure 19.5).

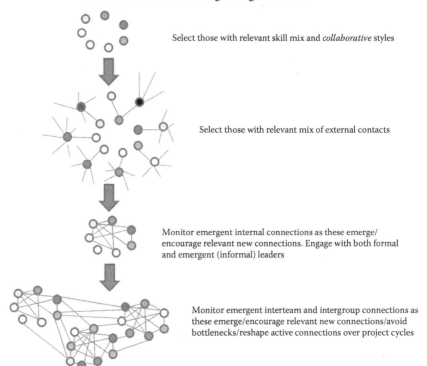

Partner innovations: Working through networks

Select those with relevant skill mix and *collaborative* styles

Select those with relevant mix of external contacts

Monitor emergent internal connections as these emerge/encourage relevant new connections. Engage with both formal and emergent (informal) leaders

Monitor emergent interteam and intergroup connections as these emerge/encourage relevant new connections/avoid bottlenecks/reshape active connections over project cycles

FIGURE 19.5
Working through the network.

An Innovation Design for the UK Innovation *Catapults*

Informal networks data can also be used to plan for how an organization will evolve flexibly, as in the design of the UK government's transport innovation network (Transport Systems Catapult, 2014). It was designed to include innovation teams with individuals selected both for their capabilities and for their personal innovation networks. The design was for monthly measures of internal team working, inter-team working, and interactions with key innovators in parent organizations. Project manager roles were to be filled by emergent leaders from within each project team, typically within the first six weeks of the project team's life. The "health" of these networks was intended as a key leading indicator of innovation success or failure. This radical design has recently been shared with all the Catapult organizations in the UK and is still under consideration.

Neil Farmer, 2014

Use of a Computer-Based Collaboration/Knowledge Platform

A modern computer-based collaboration and knowledge storage platform should be used to communicate within/across communities, and to store and provide easy access to relevant research and project information. This platform is used to exchange e-mails, attachments and voice mails, store short videos, documents, photos, and links to external information. It is cloud based and can be accessed through any computer-based desktop, laptop, or smartphone device. It supports communities in secure spaces as well as provides wide access to all relevant authorized people.

The value, flexibility, and ease of use of the platform can be illustrated in particular by the way that expert knowledge can be captured and retrieved.

Using Short Videos from Knowledge Experts

Knowledge experts (both internally and from partner organizations) will prepare short "bite-size" videos on specific practical topics by speaking informally on each topic for about five minutes. This sound track will then be used by a graphic artist to create the brief video, typically showing a few seconds of the expert talking to camera, followed by a series of graphics illustrating what the expert is talking about. Each bite-size video can then be accessed on a just-in-time basis through a catalogue on the platform that can be search by topic, keywords, expert, etc.

Neil Farmer, 2014
with input from Fuse Universal

Content on the platform varies from the very informal (tentative ideas and suggestions) through to the formal project documents, project videos, and expert videos for future reference. Metrics on individual and group usage, document access, and perceived value are generated automatically.

As content builds on the platform, just-in-time knowledge access becomes more and more powerful. It becomes a key ingredient in the innovation environment that project teams and many relevant others (internally and externally) use routinely. The platform also facilitates the creation of informal networks across time, as platform content leads to the identification of relevant contacts for new partners, projects, and other initiatives.

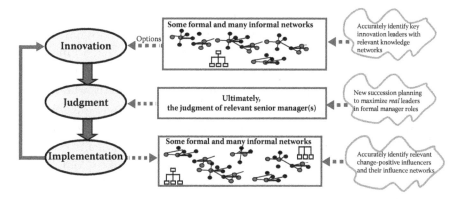

FIGURE 19.6
Synchronized formal/informal designs.

Integrating Innovation and Change Implementation into a Single Practical Model

The *networked leadership model* is based on the leadership realities in each organization and the fundamental truths outlined above. Both traditional command and control leadership and pure democratic (everyone is a leader) models are seriously flawed because they are not aligned with these workplace realities.

Although the mechanisms used will vary slightly from organization to organization, both the implementation and innovation elements will have synchronized formal/informal designs as described above (see Figure 19.6).

The additional ingredient in the model is *judgment*. For most organizations today, where *command and control* designs are commonplace, judgment on what innovations to implement will ultimately be the responsibility of relevant (often senior) managers.

Importance of Joining Up Fragmented Informal Networks

Experience with informal network analysis highlights the importance of a whole range of fragmented informal networks that exist in all organizations. When key individuals from relevant networks are brought together in teams to work on problems, opportunities, or change projects, a number of these fragmented informal networks become joined up (through discretionary effort and new opportunities, not by command and control). For example, when a number of fragmented innovation networks are brought together through key individuals in each network working together, new, larger more integrated innovation networks are formed. Fragmented expert networks can become knowledge-sharing networks by a similar process. Similarly, when key influencers work together on a change project, larger, more integrated change-positive influence networks are formed. (Since these new integrated networks are *in the know* about change developments before formal announcements are made, their comments tend to dominate informal discussions, with potential change blockers being progressively marginalized.)

Over time, we expect that more and more emergent leaders (change-positive and open-minded influencers with effective judgment) will move into formal management positions at all levels. By 2020, they will be in a majority in all leading, innovative organizations. This will greatly enhance the effectiveness of the networked leadership approach. In the meantime, formal managers (who often have an influence deficit) should exercise their judgment to put in place (at least) the innovation and change implementation elements of the networked leadership model. Much improved succession planning is just over the horizon.

Please bear in mind one of the key insights mentioned earlier: "In today's real world of work, consistently effective leadership is 'networked': effective managers will increasingly be those who work with both formal *and* informal leaders to achieve desired goals."

EXAMPLES

Examples were included in the section of this chapter entitled "How to Use the Tool."

SOFTWARE

We have no specific software recommendations.

REFERENCES

All of the informal network analysis insights and experiences in this chapter are based on practical experience of change implementation and innovation across 12 UK organizations with a total of some 26,000 employees.

In addition, the following publications contain a wide range of research findings and insights that we have drawn upon in designing the networked leadership model.

Major Change Failure Rates
- IBM Global Study. Majority of organizational change projects fail. October 14, 2008.
- Kotter, J. Leading change: Why transformation efforts fail. *Harvard Business Review*, March–April 1995.
- Nohria, N. and Beer, M. Cracking the code of change. *Harvard Business Review*, May 2000.

Informal Network Analysis
- Cross, R. and Parker, A. *The Hidden Powers of Social Networks*. Boston: Harvard Business School Press, 2004.
- Duan, L., Sheeren, E., and Weiss, L.M. Tapping the power of hidden influencers. *McKinsey Quarterly*, March 2014.
- Farmer, N. *The Invisible Organization*. Surrey: Gower, 2008.

Emergent Leaders
- Goffee, R. and Jones, G. Leading clever people. *Harvard Business Review*, 2007.
- Hlupic, V. *The Management Shift*. London: Palgrave Macmillan, 2014.
- Levine, S.R. The skills required for emergent leadership. *Credit Union Times Magazine*, April 20, 2014.

Innovation
- AIM Research. At the edge of innovation. 2008.
- Johansson, F. *The Medici Effect*. Boston: Harvard Business School Press, 2006.

70:20:10
- Jennings, C. Video: www.youtube.com/watch?v=t6WX11iqmg0.

Future of leadership
- Collins, J. *Good to Great*. New York: Random House, 2001.
- Hamil, G. *What Matter Now*. San Francisco: Jossey-Bass, 2012.

20

Solution Analysis Methods Using FAST

Frank Voehl

CONTENTS

DEFINITION

FAST is an innovative technique to develop a graphical representation showing the logical relationships between the functions of a project, product, process, or service based on the questions *How* and *Why*. In this case, it should not be confused with FAST that stands for the Fast Action Solution Team methodology created by H.J. Harrington. The two are very different in application and usage.

USER

This tool can be used by individuals or groups, but its best use is with a group of four to eight people. Cross-functional teams usually yield the best results from this activity. The FAST diagram is usually prepared in a workshop setting and led by someone with experience in preparing FAST diagrams. Input for the diagram is received from workshop participants.

The Technology Solution Analysis and Design process translates the Technology Solution Requirements into an implementable design, and finalizes the project's implementation schedule and costs. Technology components are identified and architected to provide a framework in support of the Technology Solution Build phase.

OFTEN USED IN THE FOLLOWING PHASES OF THE INNOVATIVE PROCESS

The following are the seven phases of the innovative cycle. An X after the phase name indicates that the tool/methodology is used during that specific phase.

- Creation phase X
- Value proposition phase
- Resourcing phase
- Documentation phase
- Production phase X
- Sales/delivery phase
- Performance analysis phase X

TOOL ACTIVITY BY PHASE

- Creative phase—Solutions analysis is used as a research technique and approach of identifying business needs and determining solutions to business problems.

- Production phase—During this phase, solutions analysis is used to better understand and define root cause impact related to production process and product performance.
- Performance analysis phase—During this phase, solution analysis is used to capture and analyze true impact in acceptance of the individual solution.

HOW TO USE THE TOOL

The FAST diagram or model is an excellent solutions analysis communications vehicle. It uses the verb–noun rules in function analysis to quickly create a common language, crossing all disciplines and technologies. By doing so, it allows multidisciplined team members to contribute equally and communicate with one another, while addressing the solution objectively without bias or preconceived conclusions.

FAST stands for Function Analysis System Technique, and is an evolution of the value analysis (VA) process created by Charles Bytheway. FAST permits people with different technical backgrounds to effectively communicate and resolve issues that require multidisciplined considerations for solutions to work effectively. FAST builds on VA by linking the simply expressed, verb–noun functions to describe complex systems.

FAST is not an end product or result, but rather a beginning. It describes the item or system under study, and causes the team to think through the functions that the item or system performs, forming the basis for a wide variety of subsequent approaches and analysis techniques. FAST contributes significantly to perhaps the most important phase of value engineering: function analysis. FAST is a creative stimulus to explore innovative avenues of solutions for performing functions.

With FAST, there are no right or wrong models or results. The problem should be structured until the product development team members are satisfied that the real problem is identified. After agreeing on the problem statement, the single most important output of the multidisciplined team engaged in developing a FAST model is consensus for the solution. Since the team has been charged with the responsibility of resolving the assigned problem, it is their interpretation of the FAST model that reflects the problem statement that is important (see Figure 20.1).

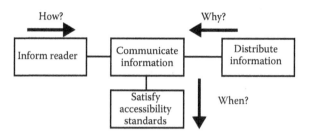

FIGURE 20.1
FAST model for a typical program.

The team members must discuss and reconfigure the FAST model until a solution consensus is reached, and all participating team members are satisfied that their concerns are expressed in the model. Once consensus has been achieved, the FAST model is complete and the team can move on to the next creative phase.

FAST differs from VA in the use of intuitive logic to determine and test function dependencies and the graphical display of the system in a function dependency diagram or model. Another major difference is in analyzing a system as a complete unit, rather than analyzing the components of a system. When studying systems, it becomes apparent that functions do not operate in a random or independent fashion. A system exists because functions form dependency links with other functions, just as components form a dependency link with other components to make the system work. The importance of the FAST approach is that it graphically displays function dependencies and creates a process to study function links while exploring solution options to develop improved systems.

There are normally two types of FAST diagrams: the technical FAST diagram and the customer FAST diagram. A technical FAST diagram is used to understand the technical aspects of a specific portion of a total product. A customer FAST diagram focuses on the aspects of a product that the customer cares about and does not delve into the technicalities, mechanics or physics of the product. A customer FAST diagram is usually applied to a total product.

FAST aids in thinking about the problem objectively and in identifying the scope of the project by showing the logical relationships between functions. By organization of the functions into a function-logic, FAST diagram enables participants to identify of all the required functions. The FAST diagram can be used to verify if, and illustrate how, a proposed solution achieves the needs of the project, and to identify unnecessary, duplicated, or missing functions. The development of a FAST diagram is a creative thought process that supports communication between team members.

The development of a FAST diagram helps teams to

- Develop a shared understanding of the project
- Identify missing functions
- Define, simplify, and clarify the problem
- Organize and understand the relationships between functions
- Identify the basic function of the project, process, or product
- Improve communication and consensus
- Stimulate creativity

HOW TO USE THIS TOOL TO CREATE A FAST DIAGRAM

Three key questions are addressed in a FAST diagram:

- How do you achieve this function?
- Why do you do this function?
- When you do this function, what other functions must you do?

Figure 20.2 illustrates how a function is expanded in How and Why directions in a FAST diagram.

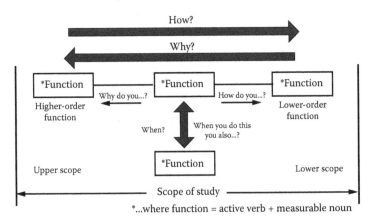

FIGURE 20.2
How a function is expanded.

EXAMPLE

Start with the functions as identified using function analysis:

- Expand the functions in the How and Why directions.
- Build along the How path by asking, "How is the function achieved?" Place the answer to the right in terms of an active verb and measurable noun.
- Test the logic in the direction of the Why path (right to left) by asking, "Why is this function undertaken?"
- When the logic does not work, identify any missing or redundant functions or adjust the order.
- To identify functions that happen at the same time, ask, "When this function is done, what else is done or caused by the function?"
- The higher-order functions (functions toward the left on the FAST diagram) describe what is being accomplished and lower-order functions (functions toward the right on the FAST diagram) describe how they are being accomplished.
- "When" does not refer to time as measured by a clock, but functions that occur together with or as a result of each other.

Consider the FAST diagram in Figure 20.3 for a mouse trap using the how and why logic as described above.

There is no *correct* FAST diagram, but there is a valid method of representing the logic in a diagram. The validity of a FAST model for a given situation is dependent on the knowledge and scope of the workshop participants. The FAST diagram aids the team in reaching consensus on their understanding of the project.

FAST is not an end product or result, but rather a beginning. It describes the item or system under study, and causes the team to think through the functions that the item or system performs, forming the basis for a wide variety of subsequent approaches and analysis techniques. FAST contributes significantly to perhaps the most important phase of value engineering: function analysis. FAST is a creative stimulus to explore innovative avenues for performing functions.

The FAST diagram or model is an excellent communications vehicle. Using the verb–noun rules in function analysis creates a common language, crossing all disciplines and technologies. It allows multidisciplined

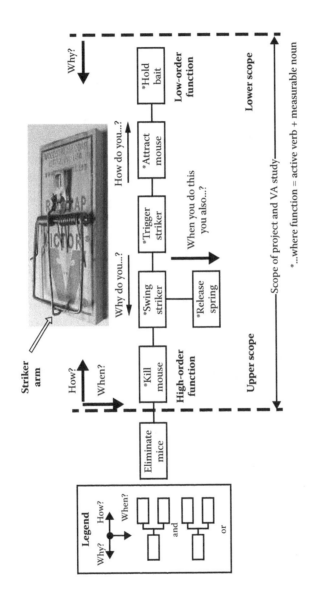

FIGURE 20.3
FAST diagram.

team members to contribute equally and communicate with one another while addressing the problem objectively without bias or preconceived conclusions.

CREATING A FAST MODEL

The FAST model has a horizontal directional orientation described as the How–Why dimension. This dimension is described in this manner because How and Why questions are asked to structure the logic of the system's functions. Starting with a function, we ask How that function is performed to develop a more specific approach. This line of questioning and thinking is read from left to right. To abstract the problem to a higher level, we ask Why is that function performed? This line of logic is read from right to left.

There is essential logic associated with the FAST How–Why directional orientation. First, when undertaking any task, it is best to start with the goals of the task, then explore methods to achieve the goals. When addressing any function on the FAST model with the question Why, the function to its left expresses the goal of that function. The question How is answered by the function on the right, and is a method to perform that function being addressed. A systems diagram starts at the beginning of the system and ends with its goal. A FAST model, reading from left to right, starts with the goal, and ends at the beginning of the *system* that will achieve that goal.

Second, changing a function on the How–Why path affects all of the functions to the right of that function. This is a domino effect that only goes one way, from left to right. Starting with any place on the FAST model, if a function is changed, the goals are still valid (functions to the left), but the method to accomplish that function, and all other functions on the right, are affected.

Finally, building the model in the How direction, or function justification, will focus the team's attention on each function element of the model. Whereas reversing the FAST model and building it in its system orientation will cause the team to leap over individual functions and focus on the system, leaving function *gaps* in the system. A good rule to remember in constructing a FAST model is to build in the How direction and test the logic in the Why direction.

The vertical orientation of the FAST model is described as the When direction. This is not part of the intuitive logic process, but it supplements intuitive thinking. When is not a time orientation, but indicates cause and effect.

Scope lines represent the boundaries of the study and are shown as two vertical lines on the FAST model. The scope lines bound the *scope of the study*, or that aspect of the problem with which the study team is concerned. The left scope line determines the basic function(s) of the study. The basic functions will always be the first function(s) to the immediate right of the left scope line. The right scope line identifies the beginning of the study and separates the input function(s) from the scope of the study.

The objective or goal of the study is called the *highest-order function*, located to the left of the basic function(s) and outside of the left scope line. Any function to the left of another function is a higher-order function. Functions to the right and outside of the right scope line represent the input side that *turn on* or initiate the subject under study and are known as lowest-order functions. Any function to the right of another function is a lower-order function and represents a method selected to carry out the function being addressed.

Those function(s) to the immediate right of the left scope line represent the purpose or mission of the product or process under study and are called basic function(s). Once determined, the basic function will not change. If the basic function fails, the product or process will lose its market value.

All functions to the right of the basic function(s) portray the conceptual approach selected to satisfy the basic function. The concept describes the method being considered, or elected, to achieve the basic function(s). The concept can represent either the current conditions (as is) or proposed approach (to be). As a general rule, it is best to create a *to be* rather than an *as is* FAST model, even if the assignment is to improve an existing product. This approach will give the product development team members an opportunity to compare the *ideal* to the *current* and help resolve how to implement the differences. Working from an as-is model will restrict the team's attention to incremental improvement opportunities. An as-is model is useful for tracing the symptoms of a problem to its root cause, and exploring ways to resolve the problem, because of the dependent relationship of functions that form the FAST model.

Any function on the How–Why logic path is a logic path function. If the functions along the Why direction lead into the basic function(s), then they are located on the major logic path. If the Why path does not lead directly to the basic function, it is a minor logic path. Changing a function on the major logic path will alter or destroy the way the basic function is performed. Changing a function on a minor logic path will disturb an independent (supporting) function that enhances the basic function. Supporting functions are usually secondary and exist to achieve the performance levels specified in the objectives or specifications of the basic functions, or because a particular approach was chosen to implement the basic function(s).

Independent functions describe an enhancement or control of a function located on the logic path. They do not depend on another function or method selected to perform that function. Independent functions are located above the logic path function(s), and are considered secondary, with respect to the scope, nature, level of the problem, and its logic path. An example of a FAST diagram for a pencil is shown in Figure 20.4.

The next step in the process is to dimension the FAST model or to associate information to its functions. FAST dimensions include, but are not limited to, responsibility, budgets, allocated target costs, estimated costs, actual costs, subsystem groupings, placing inspection and test points, manufacturing processes, positioning design reviews, and others. There are many ways to dimension a FAST model. The two popular ways are called *clustering functions* and the *sensitivity matrix*.

Clustering functions involves drawing boundaries with dotted lines around groups of functions to configure subsystems. Clustering functions is a good way to illustrate cost reduction targets and assign design-to-cost targets to new design concepts. For cost reduction, a team would develop an as-is product FAST model, cluster the functions into subsystems, allocate product cost by clustered functions, and assign target costs. During the process of creating the model, customer sensitivity functions can be identified as well as opportunities for significant cost improvements in design and production.

Following the completion of the model, the subsystems can be divided among product development teams assigned to achieve the target cost reductions. The teams can then select cost-sensitive subsystems and expand them by moving that segment of the model to a lower level of abstraction. This exposes the detail components of that assembly and their function/cost contributions.

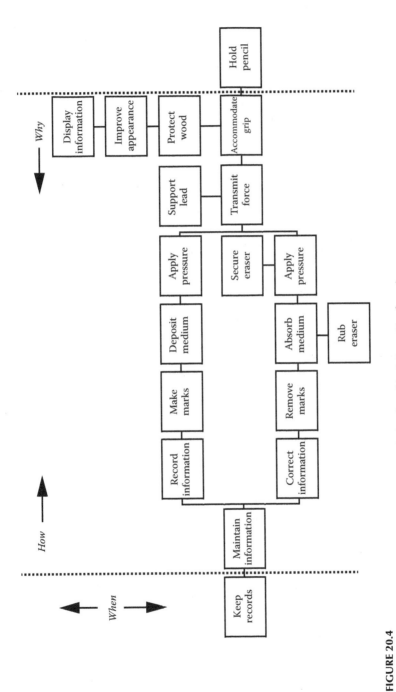

FIGURE 20.4
Fast diagram of a pencil. (Adapted from an example developed by J. Jerry Kaufman.)

Integrating Quality Function Deployment (QFD) with FAST

A powerful analysis method is created when FAST is used in conjunction with QFD. QFD enables the uses of the value analysis matrix. An example of a VA matrix for the pencil example is shown in Figure 20.5.

The steps for using these two methodologies are as follows:

1. Capture customer requirements and perform QFD product planning with the product planning matrix. Translate customer needs directly into verb–noun functions or use a second matrix to translate technical characteristics into verb–noun functions.
2. Prepare a FAST diagram and develop the product concept in conjunction with the QFD concept selection matrix. Review the verb–noun functions in the QFD matrix and assure that they are included in the FAST diagram. Revise verb–noun function descriptions if necessary to assure consistency between the QFD matrix and the FAST diagram.
3. Dimension the system in the FAST diagram into subsystems/assemblies/parts. These are generically referred to as mechanisms.
4. Develop a VA matrix at system level. The *whats* or system requirements/function in the VA matrix are derived from either a customer (versus

Customer requirements/ functions	Importance	Mechanisms						
		Lead	Eraser	Body	Paint	Band	
Make marks	30	⊙ 150					
Remove marks	20		⊙ 100				
Prevent smudges	15	○ 45		○ 45			
Support lead	5			⊙ 25			
Improve appearance	10			○ 30	○ 30	△ 10	
Accomodate grip	20			⊙ 100	△ 20		
Column weight	555	195	100	200	50	10	
Mech. weight	1.0	.351	.180	.360	.090	.018	
Mech. target cost	2.80	.98	.51	1.01	.25	.05	
Mech. actual cost	2.92	1.20	.43	.94	.10	.25	

⊙ Strong correlation weight factor = 5

○ Moderate correlation weight factor = 3

△ Weak correlation weight factor = 1

FIGURE 20.5
Example of a VA matrix.

technical) FAST diagram or by selecting those function statements that correspond to the customer needs or technical characteristics in the product planning matrix. The importance rating is derived from the product planning matrix as well.

5. Complete the VA matrix by relating the mechanisms to the customer requirements/functions and calculate the associated weight. Summarize the column weights and normalize to create mechanism weights. Allocate the target cost based on the mechanism weights. This represents the value to the customer based on the customer importance. Compare with either estimated costs based on the product concept or actual costs if available.

6. Identify high cost to value mechanisms/subsystems by comparing the mechanism target costs to the mechanism estimated/actual costs.

A product or system such as an automobile contains a great many components and would result in an extremely complex FAST model. The complexity of the process is not governed by the number of components in a product, but the level of abstraction selected to perform the analysis. With an automobile, a high level of abstraction could contain the major subsystems as the components under study, such as the power train, chassis, electrical system, passenger compartment, etc. The result of the FAST model and supporting cost analysis might then focus the team's attention on the power train for further analysis. Moving to a lower level of abstraction, the power train could then be divided into its components (engine, transmission, drive shaft, etc.) for a more detailed analysis.

In other words, the concept of decomposition is applied to a FAST model. The initial FAST model will stay at a high level of abstraction. Starting at a higher level of abstraction allows for uncluttered macro analysis of the overall problem until those key functions can be found, isolated, and the key issues identified. If a function is identified for further study, we note that with a "^" below the function box. A supporting FAST diagram is then created for that subsystem function. This process of decomposition or moving to lower levels of abstraction could be carried down several levels if appropriate.

Once high cost to value mechanisms are identified in the initial system VA matrix, the next step is to focus more attention on those mechanisms and associated functions. Dimensioning groups the functions together into those associated with a particular subsystem, assembly, or part. The

FAST diagram can be expanded into a lower level of abstraction in the area under investigation. The steps involved are as follows:

1. Use QFD to translate higher-level customer needs to subsystem technical characteristics.
2. Create FAST diagram at lower level of abstraction for targeted mechanism/subsystem.
3. Prepare a FAST diagram and develop the product concept in conjunction with the QFD concept selection matrix.
4. Dimension the system in the FAST diagram into assemblies/parts or identify the assemblies/parts needed to perform the given function.
5. Develop VA matrix at a lower level of abstraction for the targeted subsystem. The *what's* or system requirements/function in the VA matrix are derived from either a customer (versus technical) FAST diagram or by selecting those function statements that correspond to the customer needs or technical characteristics in the subsystem planning matrix.
6. Complete the VA matrix and identify high cost to value mechanisms by comparing the mechanism target costs to the mechanism estimated/actual costs.

Value Improvement Process

Performing VA or producing the FAST model and analyzing functions with the VA matrix are only the first steps in the process. The real work begins with brainstorming, developing, and analyzing potential improvements in the product. These subsequent steps are supported by the following:

- The QFD concept selection matrix is a powerful tool to evaluate various concept and design alternatives based on a set of weighted criteria that ultimately tie back to customer needs.
- Benchmarking competitors and other similar products helps to see new ways functions can be performed and breaks down some of the not-invented-here paradigms.
- Product cost and life cycle cost models support the estimating of cost for the function–cost and value analysis matrices, and aid in the evaluation of various product concepts.

- Technology evaluation leads us to new ways that basic functions can be performed in a better or less costly way. Concept development should involve people with a knowledge of new technology development and an open mind to identify how this technology might relate to product functions that need to be performed. Methods such as the theory of inventive problem solving or TRIZ (theory of inventive problem solving) are useful in this regard.
- Design for manufacturability/assembly principles provide guidance on how to better design components and assemblies that are more manufacturable and, as a result, are lower in cost.

Value analysis or function analysis provides the methods to identify the problem and to begin to define the functions that need to be performed. As we proceed in developing a FAST model, implicit in this process is developing a concept of operation for the product, which is represented by all of the lower-order functions in a FAST diagram.

Concept alternatives will be developed through brainstorming, benchmarking other products performing similar functions, and surveying and applying new technology. Since multiple concepts need to be evaluated, we want to use a higher level of abstraction for the FAST model to provide us with the greatest flexibility and a minimum level of effort. Trade studies and technical analysis will be performed to evaluate various product concepts. A concept selection matrix is a good tool to summarize a variety of different data and support making a decision about the preferred concept.

All of these steps may be iterative as a preferred concept evolves and gets more fully developed. In addition, there should be a thorough evaluation of whether all functions are needed or if there is a different way of accomplishing a function as the concept is developed to a lower level of abstraction. When a function cost or VA matrix is prepared, functions that are out of balance with their worth are identified, further challenging the team to explore different approaches.

SUMMARY

FAST is an important analysis tool, and has led to improved product designs and lower costs by

- Providing a method of communication within a product development team and achieving team consensus
- Facilitating flexibility in thinking and exploring multiple concepts
- Focusing on essential functions to fulfill product requirements
- Identifying high cost functions to explore improvements

SOFTWARE

Some commercial software available includes but is not limited to

- Creately Software: www.Creately.com. This software offers a variety of unique features to revolutionize the way you draw flowcharts. Unique features like one-click create and connect will help you save lots of time by cutting down the time taken to drag, drop, and manually connect objects. The integrated Google image search and icon finder search make it very easy to add images to your flowcharts.
- Concept Draw Solution: www.conceptdraw.com. The innovative ConceptDraw Arrows10 Technology included in ConceptDraw PRO is a powerful drawing flowcharting software that changes the way diagrams are produced, while making the drawing process easier and faster.

SUGGESTED ADDITIONAL READING

Borza, J.S. TRIZ applied to value management. TRIZCON, 2010.

Choubey, M.K. IT infrastructure and management (for the GBTU and MMTU). p. 53, 2012.

Hanik, P. and Borza, J. Enhanced brainstorming using TRIZ, 2010.

Johnston, P. FAST diagramming made easy. AASHTO Value Engineering Workshop, Minneapolis, 2013.

Kossiakoff, A. and Sweet, W.N. *Systems Engineering: Principles and Practices*. Hoboken, NJ: John Wiley & Sons, p. 266, 2011.

Moga, L.M., Ionita, I., Buhociu, F.M., Antohi, V., and Virlanuta, F.O. Introducing technical oriented fast diagrams in the projecting of the informatics systems for the management of small and medium size enterprises. *Communications of the IBIMA*, vol. 8, 2009. ISSN:1943-7765.

NDE project management (NPOESS) data exploitation website. 2008.

Robertson, S. and Robertson, J.C. *Mastering the Requirements Process*. Pearson Education, 17 mrt. 2006.

Simsion, G.C. and Witt, G.C. *Data Modeling Essentials*. 3rd Edition. Burlington, MA: Morgan Kaufman Publishers, p. 512, 2005.

System goal modelling using the i* approach in rescue. Centre HCI Design. February 27, 2003.

U.S. Department of Transportation, Office of Operations. Regional ITS architecture guidance document. July 2006.

Value methodology pocket guide. Goal/CPQ, 2008. ISBN 978-1-57681-105-4.

Wiener, R. *Journal of Object-Oriented Programming*, vol. 11, p. 68, 1998.

21

Statistical Analysis

Charles Mignosa

CONTENTS

DEFINITION

Statistical analysis is the collection, examination, summarization, manipulation, and interpretation of quantitative and qualitative data to discover its underlying causes, patterns, relationships, and trends.

USER

This tool is best used by individuals, but it can be used by small groups of two to three people.

OFTEN USED IN THE FOLLOWING PHASES OF THE INNOVATIVE PROCESS

The following are the seven phases of the innovative cycle. An X after the phase name indicates that the tool/methodology is used during that specific phase.

- Creation phase X
- Value proposition phase X
- Resourcing phase
- Documentation phase
- Production phase X
- Sales/delivery phase
- Performance analysis phase

TOOL ACTIVITY BY PHASE

- Creation phase—During this phase, it is used to conduct design experiments and to analyze data to make engineering decisions.

- Value proposition phase—During this phase, this tool is used to evaluate options and to predict results so that decisions can be made whether to continue or terminate a project or program.
- Production phase—During this phase, the tool is used to evaluate different production alternatives to maintain process control and to solve problems.

HOW TO USE THE TOOL

Statistical analysis can be used to determine five characteristics of an innovative design:

1. Establish the nature of data for analysis.
2. Estimate the performance of a larger sample or population.
3. Create a mathematical model of an innovative product or process.
4. Establish the effectiveness of a model to the actual innovation.
5. Predict variation in performance with changes in innovative elements.

Introduction

For variable data:

- Analysis of variance
  - Compares the variation in performance
  - Identifies when variation between product and process performance is significant
- t-Test
  - Used to evaluate performance differences between iteration
- Design of experiments
  - Used to evaluate the effect of independent variables on the dependent outputs
  - Used to build a model of the product or process performance
  - Used to predict future performance and stability

For attribute data:

- Quantitative statistical analysis
  - Unvaried analysis (establish the distribution against a single set of data)

- Bivariate analysis (analysis of two variables at the same time)
- Multivariable analysis (analysis of multiple variables at the same time)
- Qualitative analysis
  - Analysis of the degree to which different sets of data give a similar result

This methodology is a very complex and broad subject. It is too complex to provide the reader with all the information in the body of knowledge related to statistical analysis. As a result, this chapter will present some of the basic tools that all innovators should know and be able to use.

- Process elements, variables, and observation concepts

  Elements are the process steps, objects, or parameters on which information is collected. For example, studying the airline industry, one can identify, let us say, the top 10 airlines that become the elements in the dataset. A *variable* is a measure of interest, which is normally a process input, in-process, or output parameter, which needs to be studied and analyzed. For example, studying airlines, one may look at customer satisfaction, performance, the time of takeoff, departure time, arrival time, in-flight service, and flight load. Each piece of information collected for a variable of a process element is called an *observation*. The *observations* for a process constitute data.

- Data, scale, and sources

  Data is the first set of information that needs be collected for learning about the process performance. The data can be of attribute or variable type. The attribute types of data represent performance levels such as pass or fail, go and no-go, OK or not OK, and accept or reject. The attribute type of data is analyzed for its frequency distribution and prioritization for more data collection. The variable types of data include actual measurements such as current, voltage, speed, revolutions per minute, length, height, width, depth, distance, thickness, and many more. The variable data represent a specific value with or without decimals, which is then analyzed for its distributions and related probabilities.

  One can see that the attribute data provides information primarily about *what* is wrong, while the variable data adds *how much* or the magnitude of the problem. Normally, simple, more mature, and stable processes can use more attribute data, while new or dynamic processes must collect more variable-type data.

Scales of measurement represents the categories of information of data. There are four scales of measurement:

- Nominal
- Ordinal
- Interval
- Ratio

The *nominal* scale applies a label or name of the data that can only be categorized for analysis based on its frequency of occurrence. Such data include social security numbers, names, or problem categories. For example, analyzing data for airlines, the name of the airlines will represent the nominal scale data. Nominal data could be alpha, alpha numeric, or numeric in appearance, and it contains the minimum information.

Ordinal scale data exhibits the properties of nominal data that can be used for ranking of the process elements. For example, information collected through passenger surveys would allow you to rank airlines. Employee performance appraisal, on a scale of 1 to 5, or A, B, C, and D corresponding to the performance, would also represent ordinal scale. The ordinal data are either alpha or numeric with a relative value.

The *interval* scale data represent an additional dimension to the ordinal data when the interval between data correlate to the process performance. Interval data are always numeric. For example, GMAT scores of 600, 580, 640, and 700 for four students can be ranked from low to high, and the differences between the scores are meaningful.

The *ratio* scale represents the data where interval and ratios are meaningful. For example, consider the following on-time performance data of four airlines: 60%, 70%, 80%, and 90%. Here, both the difference (30%) between 60% and 90%, and ratio (1.5) of 90% and 60%, provide meaningful information. The variable data are of ratio scale. Table 21.1 shows an overview of various scales of measurement.

The nominal scale represents attribute data, and ratio scale represents the variable data. The nominal scale or ordinal scale data are typically the qualitative data representing limited opportunity for statistical analysis. The qualitative data can be chiefly summarized as the count or proportion of observations in each category. When data represent a count or *how many*, the data are discrete in nature.

TABLE 21.1

Overview of Various Scales of Measurement

Scale of Measurement	Description	Example	Type of Data
Nominal scale	Data collected are names or labels	Names of airlines that are operating in the USA	Qualitative data
Ordinal scale	Data exhibit the properties of nominal data and the rank of the data is meaningful	Survey result of the airline	Qualitative data
Interval scale	Data have the properties of ordinal data and the interval between observations is meaningful	Customer satisfaction score	Quantitative data
Ratio scale	Data have the properties of interval data, and the ratio of observations is meaningful	Customer service time at the airline check-in counter	Quantitative data

Source: Gupta, P. *The Six Sigma Performance Handbook.* New York: McGraw Hill, 2004.

The interval and ratio scales of measurements represent quantitative data. These are numeric values that indicate *how much* or *how many.* Quantitative data allow more opportunities for statistical analysis and arithmetic operations. Typically, quantitative data are continuous on a scale where one can specify a value or *how much,* and thus they are called *continuous data.*

Sources of data are critical aspects of process improvement. Collecting good data is important to problem solving; thus, this activity requires careful planning about what data is needed, for how long it needs to be collected, and from where it can be obtained. Sometimes data are available from a routine process, and sometimes data are collected when necessary. Sometimes part- and process-performance data are collected from external resources.

• Data accuracy

Often it is heard in meetings that data used for analysis were not accurate, or since data were not believable, no analysis was performed. The data may be inaccurate due to data entry or recording error, or incapable measuring device. In other words, data inaccuracy occurs due to either the equipment or the operator. The measuring device is unable to measure it due to its lack of calibration. The operator may not be consistent in gathering the data due to lack of training, or inability to function well due to lack of sleep. Sometimes, the data are inaccurate at first, because nobody had bothered to look at the data for so long that the operators had decided to enter expected numbers.

In any case, one can tell from the above situations that data-entry errors are easiest to find when they look excessive. One can find the data-error entry either through sorting or evaluating minimum and maximum values. Sometimes zero or decimal points are misplaced. The measurement device errors will be somewhat more difficult to determine, unless multiple devices are involved. Often, by reviewing the data, you can isolate data associated with the erroneous measuring device.

In case of *fake* data entered by the operator, the data can be isolated because the data look too good. The variation is excessively minimal, even less than normal variation. In either case of data inaccuracy, we can pinpoint the causes, and take necessary remedial actions to improve the data accuracy. You must be aware of the data errors before analyzing the data.

Statistical Thinking

In the following example, we will use the case of a Six Sigma practitioner for an airline who has been assigned the task of reducing the customer service time at the ticket counter. The first thing you would want to know is the current length of time to issue a ticket. To acquire this data, you may decide to measure the ticketing time for 50 customers using a stopwatch over several days, randomly picking different ticketing counters throughout the day to represent the customer base (see Table 21.2).

Now that you have the data, you would like to learn something from it. To analyze the data that have been collected, you should summarize the data and calculate basic statistics, such as the average time to issue a ticket, and the variation in it. After the analysis, you would make recommendations to reduce the average time to issue a ticket. As an approach, you would group the data in different ranges in minutes, construct a frequency distribution as shown in Figure 21.1, and then determine its mean value and variation.

Variation in data can be of two types: one that occurs commonly and the other that occurs rarely. Variation that occurs randomly is called *random variation*, and variation that occurs rarely is called *assignable variation*. It disturbs the normal bell-shaped curve and makes it a different-looking

TABLE 21.2

Time in Seconds for 50 Passengers
at the Airline Ticket Counters

175	198	168
185	160	175
183	166	176
189	167	177
173	159	178
172	212	178
171	161	179
169	186	162
177	172	178
180	174	178
180	188	178
181	187	213
184	200	168
188	150	163
170	199	155
165	196	181
191	163	

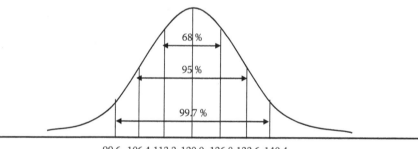

99.6 106.4 113.2 120.0 126.8 133.6 140.4

FIGURE 21.1

Frequency distribution for the time spent by passengers at the airline ticket counters.

distribution. Statistical thinking requires a clear understanding of the difference between random and assignable variations. Random variation occurs due to uncontrolled variables or circumstances that exist in the environment around the process. There are many of these uncontrolled variables that contribute a little to the process variation, are always present in the process, and are difficult to pinpoint and resolve.

The assignable variable normally does not exist, occurs rarely for a specific reason, and is easy to pinpoint for removal. For example, in case of the length of time to issue a ticket, the random variation could be due to slight variation from one passenger to the other, such as someone having lost his or her ticket, or someone forgetting to bring a driver's license. Other variations could be caused by such events as an untrained ticket agent or the computer stops working. Such circumstances cause unexpected delays that irritate customers. If the extended delays were included in the sample data, it would show excessive variation, which is not normal; thus, it will be considered an assignable variation.

Similarly, fluctuation in temperature, relative humidity, or variation in driving time to work or back to home would be examples of random variation. Arriving home late by an hour or so may occur due to a traffic jam or an accident, or a police car parked on the roadside.

BASIC STATISTICAL CONCEPTS

Basic statistics include a group of commonly used statistical tools that are used to establish statistical performance levels for a process. These measurements include mean, range, standard deviations, scatter plots, and histograms. These measures are also used to construct a bell-shaped curve or a normal distribution curve that can be used to visualize and predict process performance.

- **Summation ($\sum$ = taking the sum of)**
  The summation is the addition of a set of numbers; the result is their sum. The *numbers* to be summed (n) may be natural numbers, complex numbers, matrices, or more complicated fractions. An infinite sum is a subtle procedure known as a series.

  Because it is necessary to add (sum) numbers together, so frequently in statistics, the special symbol $\sum$ is used to indicate *take the sum of*. If there is a set of n values for a variable labeled X (n = any number of items that will be summed together), the expression $\sum_{i=1}^{n} X_i$ indicates that these n values are to be added together from the first value to the last (nth) value. Thus

$$\sum_{i=1}^{n} X_i = X_1 + X_2 + X_3 + \ldots + X_n$$

To illustrate summation notation, suppose there are five values for a variable X:

$$X_1 = 5, X_2 = 0, X_3 = -1, X_4 = -4 \text{ and } X_5 = 6$$

For these data,

$$\sum_{i=1}^{n} X_i = X_1 + X_2 + X_3 + X_4 + X_5$$

$$= 5 + 0 + (-1) + (-4) + 6 = 6$$

The following is the calculation of a second set of numbers labeled Y:

$$\sum_{i=1}^{n} Y_i = Y_1 + Y_2 + Y_3 + \ldots + Y_n$$

$$Y_1 = 1, Y_2 = 3, Y_3 = -2, Y_4 = 4 \text{ and } Y_5 = 3$$

$$\sum_{i=1}^{n} Y_i = 9$$

Sometimes, you need to sum the squared values of a variable. The sum of the squared Xs is written as

$$\sum_{i=1}^{n} X_i^2 = X_1^2 + X_2^2 + X_3^2 + X_4^2 \ldots + X_n^2$$

$$\sum_{i=1}^{n} X_i^2 = (5)^2 + (0)^2 + (-1)^2 + (-4)^2 + (6)^2 = 68$$

Note that the only thing that is different is a 2 that indicates that X is squared is added to each X value using the preceding data:

$$\left(\sum_{i=1}^{n} X_i\right)\left(\sum_{i=1}^{n} Y_i\right) = (6)(9) = 54$$

This is not the same as $\sum_{i=1}^{n} X_i Y_i$, which equals 9.

- **Mean also called *x*-bar = $\bar{x}$**

  Process mean is the most familiar and most used statistical measure. The mean typifies the expected or most likely value of a parameter. Thus, it is also most important to understand that most of the measurements cluster around the mean. The mean is calculated by adding all the data points and dividing the sum by the number of data points.

  The mean for data from a sample is denoted by *x*-bar, and the data from a population is denoted by the Greek letter μ. Each data point is represented by x_i.

  Sample mean: *x*-bar = $\Sigma X_i/n$, for $i = 1$ to n,
  where $\Sigma X_i = x_1 + x_2 + x_3 + x_4 + \dots + x_n$ and
  n = sample size.
  The Greek symbol Σ stands for *sum of*.
  Population mean: $\mu = \Sigma X_i/N$, for $i = 1$ to N,
  and
  where $\Sigma X_i = x_1 + x_2 + x_3 + x_4 + \dots + x_N$ and
  N = population size.

  Consider the data for an airline's time to issue a ticket; the mean can be calculated using Excel or any statistical software. For an MS Excel worksheet, one uses the formula as "= AVERAGE (data cell range for variable *x*)," for example: "= AVERAGE (A2:A52)."

  In case of our ticketing process example, the mean value would be 177.6 seconds.

- **Median**

  Median, similar to mean, is another measure of centrality. While the mean sums the data points and divides it by the number of observations, the median counts data points and determines the middle point in the dataset. For an odd number of observations, the middle value obtained by sorting observations in an ascending or descending order is called the *median*. For an even number of observations, the median is the average of the two middle values. Median is used when data are less variable in nature, or contain extreme values, and provides an indication of the distribution of data points. The command in MS Excel is "= MEDIAN (data cell range for variable *x*)."

In the case of our ticketing process example, the median value would be 177.5 seconds.

- **Mode**

    The mode is the value that occurs with the greatest frequency. You may encounter situations with two or more different values with the greatest frequency. In these instances, more than one mode can exist. If the data have exactly two modes, we say that the data are *bimodal*. The Excel command using MS Excel is "= MODE (data cell range for variable *x*)."

    In the case of our ticketing process example, the mode value would be 178 seconds.

- **Range**

    The range is the simplest measure of variability. It is the difference between the largest and the smallest value in the dataset. It has limitations because it is calculated using two values irrespective of the size of data. The range can be calculated by finding the maximum value [= MAX (data cell range for variable *x*)], the minimum value [= MIN (data cell range for variable *x*)], and then subtracting the two.

    In the case of our ticketing process example, the range would be 63 seconds.

- **Variance**

    Variance is a measure of inconsistency in a set of values, or the process performance, using all data values, instead of just two values used in calculating the range. It is a measure of variability that utilizes all the data. The variation is the average of the squared deviations, where deviations are the difference between the value of each observation (x_i) and the mean.

    Population (all data points) variance is denoted by the Greek symbol σ^2:

$$\sigma^2 = \Sigma(x_i - \mu)^2/N$$

Sample (subset of larger dataset, or the population) variance is denoted by s^2:

$$s^2 = \Sigma(x_i - x\text{-bar})^2/(n - 1)$$

Note: If the sample mean is divided by $n - 1$, and not n, the resulting sample variance provides an unbiased estimate of the population variance.

Variance (using MS Excel): "= VAR (data cell range for variable *x*)."

In the case of our ticketing process example, the variance would be 178.6.

Standard Deviation

Standard deviation is the square root of the variance. Thus, for the

Sample standard deviation: $s = \sqrt{s^2}$
Population standard deviation: $\sigma = \sqrt{\sigma^2}$

The command for calculating standard deviation using MS Excel is "= STDEV (data cell range for variable *x*)."

In the case of our ticketing process example, the standard deviation would be 13.4 seconds.

Correlation Coefficient

Scatter plots represent the relationship between two variables, such as experience of the employees and passenger time spent at the ticket counter, as shown in Figure 21.2. The correlation coefficient is a statistical measure of the relationship between two variables. Depending on the relationship of data

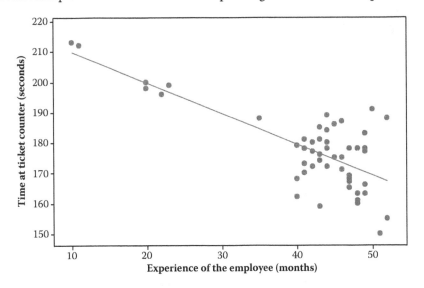

FIGURE 21.2
Scatter plot of the time at the airline ticket counters versus employee experience.

for two variables, the correlation and the correlation coefficient both can be negative or positive. The value of the correlation coefficient lies between 0 and ±1. A correlation coefficient of +1 indicates a direct positive relationship, and a value of −1 indicates a direct negative relationship. A zero value correlation coefficient implies no correlation, thus two independent variables.

The command for calculating the correlation coefficient using MS Excel is "= CORREL (data cell range for variable x, data cell range for variable y)."

In the case of our ticketing process example, the correlation coefficient would be −0.75, which is expected because more experienced employees would be able to help the passengers faster.

Histograms

Histograms are a graphical representations of the frequency distribution of data. In a histogram, the horizontal axis (X) represents the measurement range and the vertical axis (Y) represents the frequency of occurrence. Histograms are one of the most frequently used graphical tools for analyzing the variable data. Histograms are like bar charts, except that they have a continuous X-scale versus the separate and unrelated bars.

Histograms display central tendency and dispersion of the data set. If someone plots the limits around the process mean, one can tell what

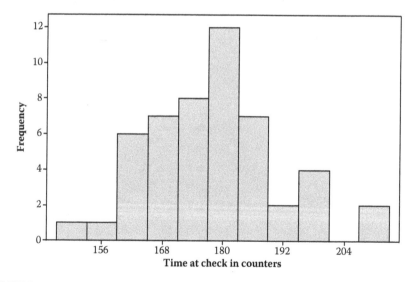

FIGURE 21.3

Histogram of time spent by passengers at ticket counter. (From Minitab Inc. MINITAB Statistical Software, Release 14 for Windows, State College, Pennsylvania, 2003. MINITAB® is a registered trademark of Minitab Inc.)

percent of the data is beyond the limits. In business, this relates to being able to determine the acceptable and reject rates of a process. One can also fit some known distributions visually or by using a software to estimate the probability of producing a good product or a bad product. By being able to determine the expected process performance, one can initiate a preventive action if necessary, rather than wait for it to happen, and then correct it.

Figure 21.3 shows the histogram with central tendency and the dispersion for the time spent by the passengers at the airline ticket counters.

BASIC PROBABILITY CONCEPTS

Probability is a measure of the likelihood that some event will occur. The probability of the occurrence of any event (A) is defined as:

$$P(A) = \text{Number of times event A occurs/can occur}$$
Number of times the experiment was conducted/total possibilities

The probability of any event's occurrence ranges from 0 to 1. For example, we want to know the probability of finding 3 when we toss a fair die once. In this case, we know we can get six different values and 3 can come only once. Hence, the probability of finding 3 in the toss of a die would be

$$P(3) = 1/6 \text{ or } .17$$

Probability Theory

Most of the statistical concepts are based on probability theory. One of the ways to demonstrate probability theory is through the games we play, like using dice or cards. Let us start with the dice example. A single die has the following specifications:

a. Has only six sides.
b. Only one side can be up at a time.
c. Each of the six sides has the same probability of occurrence.
d. The probability of any one side coming up is 1 in 6 or .1667.
e. Each of the six sides has a different number on them that range from 1 to 6.

The probability of a single event (in the case of a die, the same number) is P(A) if A is independent of any other event. [Example: P(A) could be the probability (P) of 1 (A) coming up on one die or 16.67%.]

Now if you add a second die (B), its probability of throwing 1 is the same as die (A). But if you throw the two dice together, the sum of the numbers is dependent on the probability of a specific number coming up on die A or P(A) and a specific number coming up on die B or P(B). This is the probability of both occurring at the same time. This joint probability of A and B occurring at the same time is multiplicative by nature and is equal to P(A) × P(B), or the probability of rolling any given combination with two dice is = 1/6 × 1/6 = .1667 × .1667 = .0278 or 2.7896%.

There is only one combination that can produce a 2 (i.e., both dice come up with 1s) or 12 (i.e., both dice come up with 6s). As a result, we can say they are mutually exclusive of each other, which means they cannot occur at the same time. Any time that two events are mutually exclusive, the probability of the event occurring is calculated by summing their individual probabilities P(A or B) = P(A) + P(B) or .278 + .278 = .556 or 6.56%. In other words, the probability of rolling a number from 3 to 11 will be 94.44% every time you throw the dice. Table 21.3 shows the total combinations of rolling two dice.

There is a possibility of rolling 36 different combinations; each of them have a probability of occurrence of P(A) × P(B) = .1667 × .1667 = .0278.

By analyzing Table 21.4, we can see that there are five combinations that can result in a total of 6 (see Table 21.5).

Because all five events are mutually exclusive, their total effect is the sum of all five individual probabilities or 13.9%. Table 21.5 shows the probability of rolling any specific number with two dice.

TABLE 21.3

Dice Combinations

	1	2	3	4	5	6
1	2	3	4	5	6	7
2	3	4	5	6	7	8
3	4	5	6	7	8	9
4	5	6	7	8	9	10
5	6	7	8	9	10	11
6	7	8	9	10	11	12

TABLE 21.4

Probability of Rolling a 6

Die 1 + Die 2	Sum	Probability of Occurrence
5 + 1	6	.0278
4 + 2	6	.0278
3 + 3	6	.0278
2 + 4	6	.0278
1 + 5	6	.0278
	Total	.139

TABLE 21.5

Two-Dice Combination Probabilities

Value	Combinations	Probability
1	0	0
2	1	.0278
3	2	.0556
4	3	.0833
5	4	.1111
6	5	.1389
7	6	.1667
8	5	.1389
9	4	.1111
10	3	.0833
11	2	.0556
12	11	.0278
	Total	1.000

Now let us look at a deck of 52 playing cards. The probability of selecting a specific card out of the deck without looking is 1/52 or .0192. To calculate the probability of being dealt a royal flush (five cards A, K, Q, J, and 10 all of the same suit), see Table 21.6.

In this case, the events are not mutually exclusive because to have a royal flush, all cards have to be the same suit and be from an ace to a 10 with no one value duplicated.

In this case, the total probability of having a run from ace to 10 in the same suit (a royal flush) is calculated by multiplying the probability of the individual five cards together or

$$P(1) \times P(2) \times P(3) \times P(4) \times P(5) = .3846 \times .0784$$
$$\times .06 \times .0408 \times .0208 = .000154\%$$

TABLE 21.6

Dealing Five Cards

Card	Probability
1st Card—5/13	.3846
2nd Card—4/51	.0784
3rd Card—3/50	.0600
4th Card—2/49	.0408
5th Card—1/48	.0208

The probability of being dealt a royal flush is 649,739 to 1. You will note that the probability for the first card is 5/13 (.3846) because it can come from any of the four suits and can be any of the five cards A, K, Q, J, or 10. After the first card, the other four cards must come from the same suit that the first card came from.

Mutually Exclusive Events

Two events are called *mutually exclusive* when they cannot happen at the same time. For example, a 25-year-old individual and becoming president of the United States are mutually exclusive because according to the U.S. Constitution, an individual less than 35 years of age cannot become a U.S. president. When the two events (A and B) are mutually exclusive, the probability of occurrence of either of the events is

$$P(A \cup B) = P(A) + P(B)$$

For example, if we want to know the probability of finding 1 or 2 when we toss the die twice, it shall be

$$P(1 \cup 2) = P(1) + P(2)$$

We know from the earlier example that the probability of finding any number from 1 to 6 in the toss of a die is 1/6 and hence

$$P(1 \cup 2) = 1/6 + 1/6 \text{ or } 1/3 \text{ or } .33$$

- **Complementary events**
  When events have only two possible outcomes, they are called *complementary events*. For example, when we toss a fair coin, the two possible outcomes are head or tail.

Probability of occurrence of a complementary event = $1 - P(A)$.

- **Independent events**

  Two events are independent if the outcome of one event does not depend on the outcome of the other event. For example, if we toss a coin twice, the second outcome shall not depend on the first outcome; that is, if we get heads in the first toss, we may get heads again in the second toss.

- **Conditional events**

  When the two events depend on each other, that is, the outcome of the second event shall be a function of what happened in the first event, this is called a *conditional event*. For example, when we draw two cards from a deck of 52 cards and want to know the probability of finding an Ace as the second card. Now, whether the second card shall be an Ace depends on what was found in the first card drawn.

  The conditional probability of an event (A) given that event B has occurred is expressed as

$$P(A/B) = \frac{P(A \cap B)}{P(B)}$$

Let us apply this concept to the problem of drawing the cards:
- Probability of finding an Ace as the first card, that is, $P(\text{Ace } 1) = 4/52$.
- Probability of finding Aces as the first two cards, that is, $P(\text{Ace } 1 \text{ and Ace } 2) = 4/52 \times 3/51$.
- Probability of finding an Ace as the second card given the first card was an Ace, that is, $P(\text{Ace } 1/\text{Ace } 2) = P(\text{Ace } 1 \text{ and Ace } 2)/P(\text{Ace } 1) = (4/52 \times 3/51)/(4/51) = 3/51$ or .06.

PROCESS MEASUREMENTS

The basic tenet of Six Sigma is *measure what we value*. It is understood that if a process adds value, and needs to reduce its waste, one must be able to measure it. The initial measure of waste is the defect per unit (DPU) at the product or the process output level, the defects per million opportunities (DPMO) at the process, sigma, and business levels. Typically, DPU is measured first as it is easier and it affects the process customer directly.

The higher the DPU at a process, the more the process is likely to cause customer dissatisfaction, and producing waste in the process. Once the DPU is known, the DPMO is needed to relate to the opportunities for mistakes to occur.

The DPMO is also used for benchmarking various processes in an organization in dealing with a variety of products. Thus, the complexity of the product and process affects the number of opportunities for things to go wrong. Another way one can look at DPMO is as a way to normalize the DPU measure based on the complexity of the operation. The probability of producing defects is directly related to the number of opportunities in the process. Thus, the higher the number of opportunities, the more defects are likely to occur. For example, if a company manufactures a smaller product like a cell phone with about, let us say, 300 opportunities, and a large system with 5000 opportunities, the system with 5000 opportunities is going to have a higher number of defects.

In other words, in order to reduce the number of defects in the process or a product, we must try to reduce the number of opportunities or simplify it. Thus, if you are in the market for a product or a device, the device that looks simpler is also going to be more reliable to use, as it is less likely to fail.

Unit: A discrete output of a process is called a unit. The unit represents the deliverable of a service, or the output of a process, or a product of a company.

Defect: Any attribute of the unit that fails to meet the expected target. The characteristic that does not meet customer requirements away from the target and beyond the economic tolerance can become a defect. Thus, one can see that a defect(s) can be caused by excessive variation in the process limits of customer expectations with respect to the process capability.

Opportunity for error: An activity, task, component, item, piece of information, or event that presents an opportunity for making a mistake(s) or producing a defect.

If one examines the process of a ticket agent for an airline, one can see first what the agent needs to do, which may include gathering passengers' needs, verifying their identity, using the airline's reservation system (this may have some critical entries), printing the ticket, weighing the luggage, checking in the baggage, attaching a tag to the baggage, putting the baggage on the conveyor, and delivering the boarding pass and baggage tags to the passenger. The above steps add up to 10 opportunities for making an error.

To identify opportunities, one can identify process inputs, outputs, and in-process activities and variables. For a product, the number of components can constitute the number of opportunities. One of the caveats is to count opportunities that matter, not the one that never goes wrong, cause to make any mistake, or the application. For example, in mechanical assembly, counting every single washer may increase the number of opportunities. We must remember that our fundamental objective is to reduce the number of opportunities, not to increase them. However, depending on the application, some washers could become a critical opportunity for a defect to occur, such as in the case of airplanes, or space vehicles.

To determine the sigma level, one first needs to know the process performance in terms of defects, errors, or yield. DPU is used to measure the process performance.

DPU = total number of defects or errors/total number of units verified

DPMO = DPU * 1,000,000/number of opportunities for error in a unit

To determine the first pass yield, that is, the yield of a process without any repair or rework, the following formula is used:

$$\text{First pass yield (FPY)} = e^{-DPU}$$

The command for calculating FPY using MS Excel is "= EXP (–DPU)."

For example, let us calculate the DPU for the airline's service. In a day, the ticket agents serve 250 customers. Even though there may not be a measure in place currently at the process, let us say we have 15 customers left very dissatisfied. Of 15 unsatisfied customers, the ticket agents made 25 mistakes, or could not address customer issues.

$$DPU = 25\,\text{mistakes}/250 = 1$$
$$DPMO = (.1(DPU)/10\,\text{opportunities}) * 1,000,000$$
$$= 100,000$$

Now, one can determine the sigma level from a conversion table or by using a software program. Since a three sigma level corresponds to

66,810 PPM, one can for sure say that the sigma level for the ticket agent is less than three sigma. Accordingly, the corresponding sigma level is 2.78.

It has been observed that sometimes companies do not track defects even though it is critical to have the right measures. Understanding the real-world situations, one can still use the yield to calculate the DPU and the corresponding sigma level. The formula for calculating the DPU from the process yield is as follows:

$$\text{Calculated DPU} = -(LN(Yield\%/100)),$$

where LN represents natural logarithm.

The command for using MS Excel is "= (−(LN(Yield/100))."

Using Process Capability Indices

The purpose of knowledge about the capability of a process to meet its requirements is pretty obvious. It provides answers to two questions: Can a process as currently designed, equipped, and operated meet its requirements? If not, what proportion of the output is expected to be defective? The indices are just one of the tools for management to use to facilitate the investigation, analysis, and corrective actions necessary to bring the process capability up to the point of filling all its requirements.

There are a variety of conditions or stages of a process as it moves from design to full production, and there are a variety of depths of analysis that may be applied depending on many factors. A number of different terms are used to identify a given analysis in a given stage of the process. They might be called process capability studies, process performance evaluations, process improvement programs, process optimization, etc. What they are called is not too important; however, what is important is to realize that nothing is constant. Even if a process is capable today, it may not be tomorrow. Thus, there is needed a never-ending effort to maintain whatever level of proficiency has been achieved and to continuously strive to reduce the natural variation of all processes so that the levels of performance to requirements are improved and the goal of zero defects is sought.

It is necessary to understand that *process capability* as expressed by some index is not an *absolute* but rather an attempt to quantify and compare the degree to which the various processes can be expected to meet their requirements. For instance, the numerical value of a process capability index can differ for a number of reasons. If the index was derived from control chart samples of past history, it might not be the same as one derived from samples taken today. If the sample sizes and frequency of samples were to be changed, one might get a different value of the standard deviation of the process, thus changing the process capability index. To be more specific, if samples within a subgroup are taken very close together, the R value will tend to be smaller than if the samples in the subgroup are taken over a longer period. The latter will increase the value of R, thus increasing the size of the estimated standard deviation. That in turn broadens the control limits and decreases the process capability index. The point of this discourse is to help you realize the fallibility of process capability indices. They are very valuable tools, but they must be used with judgment and an understanding of the conditions under which they were derived.

Different Type of Partitioning

Partitioning can help you decide how to spend your time, money, and resources. Vilfredo Pareto, an Italian economist (1848–1923), studied the distribution of wealth of the world and found a significant pattern. A few people had most of the money. For example, in the United States in 1996, 17% of the people controlled 77% of the publicly held stocks. This way of looking at the distribution of wealth leads to a special pattern called the Pareto principle.

Definition: The Pareto principle says that about 80% of the results come from about 20% of the causes.

Roughly 80% of the goals scored in the National Hockey League are scored by 20% of the players. Using this theory in your world, 20% of your customers are causing 80% of the complaints. If you identify this vital few, then you can find ways to make your customers happy. Application of the Pareto principle is a major tool for determining when and where to make improvements.

Pareto diagrams are used to analyze data from a new perspective, focusing attention on problems in priority order, comparing data changes during different time periods and for providing a basis for showing the cumulative effect of a problem. A Pareto diagram provides an easy-to-see graphical comparison. Geometric shapes are easier to compare than verbal descriptions. A Pareto chart gives you a starting point to examine a situation.

The example in Figure 21.4 shows that the first two issues (20%) cause almost 80% of your golf scoring problems. If you can solve your slice and keep from looking up, you will have made a significant improvement in your golf game and reduce your score significantly.

Partitioning or stratifying can save you a great deal of time and effort. Using a simple tool such as a Pareto chart can help you focus your resources on the areas where they will have the most benefit.

Be careful to make your samples random or you will fail to see patterns that you should and will see patterns that really do not exist.

Stratification is the process of arranging things that are similar. You identify similar things by type. The collecting of data for each group or strata allows you to compare. If a distribution is not normal, stratification may be misleading!

You can be more successful in making a decision with numbers by using proper sampling techniques. Remember:

- A sample is representative of an entire population only when it is random.
- Choosing a random sample allows you to make generalizations about the entire population based on information gathered from the sample.
- Use a Pareto chart to stratify your data.
- If you do not understand your purpose for collecting data, you may collect the wrong data.
- Determine which sampling technique is appropriate.

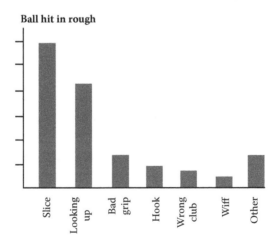

FIGURE 21.4
Pareto chart of golfer's problems.

FIGURE 21.5
Card game puzzle.

There are two different kinds of errors. One error is failing to change when something should be changed. The other is changing something that does not need to be changed.

Just for Fun

Dan and Carolyn played games all evening. They were playing for $10 per game. At the end of the evening, Dan had lost $30. He had won three games. How many games did they play? How many games did Carolyn win? (See Figure 21.5.)

The answer is, they played nine games. Dan has won three games and Carolyn has won six games.

PROCESS CAPABILITY STUDY—STEPS

Set forth below is a typical series of steps that can be used when undertaking a process capability study.

- Step 1. Preparation
  Become familiar with the process to be studied. What physical shop conditions exist (tools, machines, materials, manpower, etc.)? What are the specification limits? What variables are to be studied? What historical data exist? What data needs to be collected?

- Step 2. Data collection
    a. Select the size and sequence of the subgroups to be measured.

       Typically, a rational subgroup of four or five pieces is sufficient. These need to be taken from a single stream of production over a short interval of time (normally sequentially produced). They should represent, as nearly as possible, production under one set of conditions.
    b. Decide the frequency of the subgroups to be chosen. Subgroups taken very frequently will reflect variations that occur with little external influence, while those taken over a 24-hour period will reflect the effect that operator change, tool wear, etc., have on production. The objective is to detect all the changes the process will normally undergo.

Statistical Process Control

Definition: Statistical process control (SPC) is using data for controlling processes, making outputs of products or services predictable. It is a mathematical approach to understanding and managing activities. It includes three of the basic statistical quality tools: design of experiments, control charts, and characterization.

The goal of SPC is defect or error prevention rather than just detection (after the fact). SPC involves the use of statistical probability analysis to evaluate and monitor process performance. These techniques help in the identification of *special* and *common* causes that affect the process performance. Common causes are inherent to the process. They are caused by the variations and interactions of the people, machines, raw materials, etc., that make up the process. Special causes are due to some abnormality that prevents operational stability.

Understanding this distinction and how to apply the principles of SPC makes it possible to remove *special* causes to achieve a stable condition and to determine ways in which the continued reduction of variation can be achieved. SPC helps the organization understand its processes and obtain a predictable performance level needed to meet the never-ending demand for higher quality.

SPC is most effective if it is part of a total quality management system, but it can be used in conjunction with or separately from almost any organizational improvement process like reengineering, redesign, or Six Sigma.

To understand SPC, there are eight essential tools an individual or team should become familiar with. These tools are

1. Data collection
2. Sampling
3. Frequency distribution
4. Stratification
5. Variable control charts
6. Attribute control charts
7. Scatter diagrams
8. Process capability

Laying the foundations for SPC is a management task that includes establishing "quality awareness" throughout the entire organization. W. Edwards Deming stated, "Everyone in the company must learn the rudiments of the statistical control of quality, not just to solve a problem, but as a plan of knowledge by which to find problems and the causes thereof. It will not suffice to have some brilliant successes here and there."

SPC is a term used generally to describe a concept and methodology using proven statistical analyses to

- Determine whether a repetitive activity is in a predictable state
- Enhance the ability to manage the activity in the predictable state once it is attained

SPC should operate in a holistic context involving each person in the organization. It is a vehicle that responds to the quality themes of Deming, Juran, Crosby, Harrington, and Feigenbaum. Understanding how a system works allows us to deal with reality and to make changes in the system so the system will be as effective as we can make it.

Since systems are apt to be controlled by management, management must develop and espouse the critical concepts required for statistical methods to be useful and therefore maximize organizational performance. These critical concepts are outlined below. Without these, there can be no effective process or system approach to quality control.

- Management must strive for the prevention of defects instead of the detection of defects.
- Left to natural forces, there is no such thing as unchanging operational performance. (There are only two states in nature: operational performance improvement and operational performance deterioration.)
- A continual and determined effort to improve quality must be incorporated into management practices. The mentality of acceptable quality level (AQL) can no longer be accepted.

  AQL is a term used to quantify the percentage or proportion of defects or defectives that is considered satisfactory quality performance for a process or product. However, making decisions on quality using attribute sampling and AQLs has a tendency to maximize cost in the long run, and is therefore not recommended for use.
- The importance of quality cannot be delegated, understated, or assumed. It can be understood only through the training of all employees from executive officers to the active workforce.
- Process or system control is the only way in which quality can be defined in predictable terms. Samples taken from lots of batches produced under unknown conditions cannot provide accurate predictable information about the quality of those lots of batches.
- All resources used in systems or processes must come from stable sources. Statistical evidence of quality must be provided with each resource in order to have confidence in the quality of the process output. This knowledge is needed in order to maximize the output quality of the process.
- Only statistical analysis, using proven statistical techniques, can provide evidence of the necessary quality that is acceptable by customers. The most successful analysis of data is one in which graphical techniques, based on applicable theory, are used to establish whether the process output comes from a stable source or not and whether it is within specification requirements.
- Special or assignable causes are responsible for nonstable conditions. These causes are usually operation oriented and can be corrected at the operator level. (Dr. Deming estimated that 15% of all causes are *special* causes.)
- Common or chance causes are those that are the result of process variations. Corrections or reduction in process variation require

physical changes to the process or system. (Dr. Deming estimated that 85% of all causes are *common* causes.)

- The customers' demand for quality requires that *no* defectives be delivered and only statistical evidence be accepted as proof of the quality delivered.
- Vendors and suppliers, both internal and external, will need to implement process control programs. The responsibility for educating vendors and suppliers is that of the customer. Only if vendors are trained can they be expected to learn what is required in quality and supporting evidence.

Let us elaborate on these points. The basic concept of SPC, that is, using statistical signals to indicate the need for actions to improve performance or output, is almost universal. It can be applied to any area where work is done, where outputs exhibit variation, and where there is a desire to make improvements in that work or output.

The desire for improvement goes hand in hand with a strategy that emphasizes prevention rather than detection. After-the-fact inspection is not cost-efficient and is unreliable because at the point of inspection, wasteful and unreliable production has already occurred. It is much more effective to avoid waste by not producing goods that will be unusable in the first place—a strategy of prevention. This strategy is exemplified in the slogan "Do it right every time."

An understanding is required of the elements of an SPC system in order to effectively use SPC:

- A process control system is essentially a feedback system. There are several elements in the system.
- The process is the aggregate of the people, equipment, materials, methods, and environment that work together to produce goods or services (output).
- The ultimate performance of the process depends on how that process has been designed, how it has been constructed, and how it is being operated.
- The system is useful only if feedback from it is used to improve the performance of the process.

Feedback or information on the output from the process, when gathered and correctly interpreted, indicates whether the process or its output needs to be changed. If action is deemed necessary, it must not only be appropriate, but also must be timely in order to take full advantage of the feedback (data) gathered on the operation of the process.

There is a difference between action on the process and action on the output. Action *on the process* is future oriented; that is, appropriate and timely actions *prevent* the production of output, which is not within specification. This type of action could theoretically be taken on any of the components integral to the system. The effects of any action taken should be monitored and further actions taken if analysis of feedback indicates that a further change in the process is required.

Action *on the output* is after the fact; that is, it involves *detecting* already-produced output, which is outside of specification. Acceptance sampling is too late to have any effect on the inherent quality of products already produced. If output does not consistently meet requirements needed by customers, it may be necessary to institute a costly sorting, reworking, or scrapping operation for any items not conforming to specifications. This damaging effort to a quality and productivity program must continue until the process has been corrected or until requirement specifications for the product have been modified. Obviously, inspection of output only is a poor substitute for doing it right the first time.

- Variation

    To effectively use the data generated from observing a process, it is necessary to understand the concept of *variability*. Variation is a part of nature—no two things, products, or characteristics are exactly alike. Uniformity does not exist because any process contains many sources of variability. As the tolerance requirements of a process get more stringent, the harder it is to control the variation that exists naturally within that process.

    Three types of variation can be described: periodic, trend, or independent. In addition, there can be combinations of these three types. Consequently, time periods and conditions for which feedback (data, measurements) is obtained affect the amount of total variation that is represented in the feedback. Examples of causes of variation that occur over long periods are tool/machine wear, changes in environment, materials aging, and changes in methods or procedures.

    To manage any process, to reduce variation, requires tracking a variation back to its sources and treating the variation there. To do this requires an understanding of the differences between *special* (assignable, operational) causes and *common* (system, natural) causes.

- *Special* or *assignable causes* of variation are all those that are not inherent in the system but rather occur as a result of operating the system.

  These operational causes most usually occur as a result of the skill level of the operators, the source of materials that they use, the tools used, the procedures that are followed, the operating condition of the equipment or machines, or the physical environment when the process is operated.

  Operational causes of variation usually require some *local* action. Unless and until all these operational causes of variation are identified and corrected, they will continue to have an unpredictable effect on the process output, thereby preventing the process from being brought into a state of statistical control.

- *Common causes* refer to the sources of variation within a process that is in statistical control. These causes occur randomly during the operation of the process and cannot be isolated as unique, definable causes. Simple statistical techniques will indicate the extent of the common causes of variation. It is usually management's job to improve on these common causes, although other people directly connected with the process are in a better position to identify these causes and alert management. Improving common causes requires deliberate, planned actions to physically change the process by altering the resources used to establish or operate the process.

Usually, only a relatively small percentage of most process problems can be corrected by those most closely connected with the operation. The majority of correctable causes can only be corrected by management action on the system. Drs. W. Edwards Deming and Joseph M. Juran have observed that the proportion of problems correctable by online workers is about 15%. They say that management action is required to correct the other 85% of the problems. This, of course, varies with the nature of the process and the degree of responsibility for the system that is allocated to the online workers. What is important is not the percent division between management and workers but rather the recognition of the fact that the system accounts for a large percentage of the problems. Only by statistical analysis can the problems be sorted out and appropriate action taken by those responsible for such system changes.

These system causes of variation result in defects when variation exceeds the required specified tolerances. Correction of these defects requires continued refinement of the physical operating nature of the process. This could come from refinements in technological precision; reduction in variation of materials used; the enhancement of operator skills through training; the description of operating procedures in clear, accurate language; and control of environmental conditions.

- Measuring process outputs

  The measurement of process output is also a possible source of error in describing process behavior. Measuring device calibration must be checked to prevent biases in describing process behavior. Measurement procedures must also be verified for consistency in measurement. Errors in measurement are special causes that can mislead the portrayal of process behavior.

  The goal of an SPC system is to enable us to make sound decisions about actions affecting the process. Not to be forgotten, however, is the economic impact of such decisions. You may not need to have such close tolerances as you are striving to achieve.

  A process is considered to be in statistical control when the only sources of variation are common causes. Dr. Deming stated that, "A state of statistical control is not a natural state for a manufacturing process. It is instead an achievement, arrived at by elimination, one by one, by determined effort, of special causes of excessive variation."

  The first function, then, of a process control system is to provide feedback on the behavior of a process. Process behavior is the description of a process where a sequence of samples is taken and a particular characteristic (attribute or variable) is measured and plotted on an appropriate statistical chart. This feedback provides information leading to inferences about process stability and process capability.

  - *Process stability* exists when a predictable pattern of statistically stable behavior is demonstrated by a sequence of observations made and plotted on appropriate charts with all interpretation rules being satisfied.
  - *Inherent* process capability is the range of variation that will occur from the predictable pattern of a stable process.

- *Operational* process capability is determined by the manner in which the process is operated in respect to how this predictable pattern meets specification requirements.
- Process capability analysis

     As stated earlier, there will be a number of different conditions under which there will be a need to perform a process capability analysis. Typically, manufacturing processes undergo several stages of development before they are ready for *full production*. There exists a need to attempt to determine the capability of the various processes during these development stages. Unfortunately, the earlier the stage, the less hard data are available. During the design stage, any attempt to assure process capability will need to be done using estimates only. Because of the use of estimates, there is a need to provide for a greater margin of error; thus, the use of a CP = 1.67 (±5σ) as a target is desirable. As development proceeds and the process moves into pilot-line production, there will need to be a concern about the effect on process capability of such elements as the tools, machines, setups, and the operators. While capability analysis during this stage can be computed from hard data, it is still difficult to predict what the *full-blown* production capability will be.

     The most frequent use of the process capability measurements and indices occurs for processes that are already ongoing. The need for the study may be brought about by such things as customer complaints, poor yields, design or production methods changes, or just the recognition of the value of the search for constant quality improvement. Processes may be meeting all their requirements; however, through process optimization, better quality products and reduced costs can be achieved.

     Most things in life are not good or bad by themselves. Good or bad is relative to some frame of reference. Is 105°F good or bad? If you are painting your house and the temperature is 105°F, that is bad. If you are in a sauna at the gym, 105°F is good. Process variation must be compared with what is desired. Requirements are determined by the customer and how they use what you provide them. Simply stated, compare what you *got* with what you *want*. This comparison is called process capability. This chapter develops a measure of *wants* and *gots*. The measure is called Cp and its big brother Cpk. As usual, the math is less important than the concept, but often we learn the concept by doing the math.

Customers are not always the same. It is hard to hit a target unless you know what the target is. For instance, if you are cooking a steak for someone, they will be much more satisfied if you cook it the way they like it.

You must find out what is most important to your customers. First determine who your customers are. Then you need to understand their requirements. Finally, it is time to determine how well you are meeting their requirements. In other words, answer these questions:

- Who are my customers?
- What do they want?
- What are they getting?
- How do I measure what they are getting?

In today's markets, consistency of products and services is being called for. Process capability is a good way of understanding the results of all of your hard work. Process capability measures the ability of your process to meet your customers' requirements (specifications).

Definition: A *process* is everything that works together to produce a product or service.

Most organizations have only 8 to 12 processes. However, each of you also has 8 to 12 processes in your jobs. Outside of your job, you also have another different 8 to 12 processes that are important to you.

This may sound confusing but it is important to understand. Think about your daily life for a minute. You have many processes. For instance, one process you have is how you pay your bills. Another is how you get to work. Take a few minutes and list the 8 to 12 most important processes that make up your daily life.

Process capability is the ability to satisfy your internal or external customers' needs. Who is your customer for your bill-paying process? The customer is the company you are paying. You have many steps in your process, such as earning money and mailing the payment. Your process capability index would measure if you paid the bills on time.

Process capability rates the ability of your system to meet your customers' need. It is a statistical measure of the relationship between variations in your process and the specification. You cannot determine process capability until the process is stable.

- Cp ratio

    By applying a few formulas, you can easily tell whether your process is capable of meeting specifications.

Definition: Process capability (Cp) compares what you are currently doing with what your customer wants.

A histogram shows what you do. Your customers' wants are your goalposts. You need to kick the ball through the goalposts.

If the variation of what you do is less than what the customer wants, your Cp is greater than 1. The bigger the number for the Cp index, the better off you are. If the Cp of your process is equal to 1, the variation of your process is the same as your customers' specification. On the average, your process will produce something the customer does not want about three times out of a thousand. If the Cp index of your process is less than 1, it is unable to meet the customers' requirements. Not meeting the customers' need is a bad thing.

The process capability index is the ratio of the spread in a process (allowed by the specification limits) to the actual amount of variation in the process. You can use the simple equation in Figure 21.6 to calculate Cp.

The upper and lower specification limits are what the customer wants. Another way to think about it is how much variation the customer can tolerate. Your current level of variation is what your customer receives. This is what you do or produce. As you recall, the measure of variation is the standard deviation or s. Measuring 6 s includes 99.7% of what you do. Divide what you do into what the customer wants. Start with the upper specification limit (USL) and subtract the lower specification limit (LSL). Divide the result of the difference by 6 s. The resulting total is the process capability or Cp.

Here is an example of a water-testing process. The specification for the process is 6 ml to 10 ml. The standard deviation is 1 ml. The Cp index is 0.67 (see Figure 21.7).

A process with a capability index less than 1 is not capable of meeting the specification. The spread of the distribution minus the

$$Cp = \frac{USL - LSL}{6\sigma}$$

FIGURE 21.6
Process capability (Cp) equation.

$$Cp = \frac{10\text{ ml} - 6\text{ ml}}{6\text{ ml}} = \frac{4\text{ ml}}{6\text{ ml}} = 0.67$$

FIGURE 21.7
Water test example.

amount of variability inherent in the process is a greater amount of variability than the specifications will allow. The customer will consistently be receiving out of specification service or product. To avoid delivering bad service or product, you need an extensive and expensive inspection system.

We know that sounds confusing, so we will give you an example that happened to one of us. We were moving to a new home in the country. There were two ways to the house; however, the shortest way crossed an old wooden bridge like that shown in Figure 21.8.

I was riding my motorcycle, my wife was driving our car, and the moving company was driving the moving van. The moving van was 12 feet wide (see Figure 21.9). When it reached the bridge, it was unable to cross because the Cp was <1. The Cp ratio tells us our

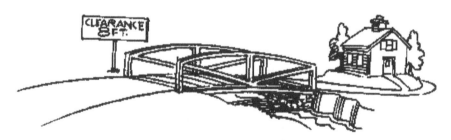

FIGURE 21.8
Bridge to cross.

FIGURE 21.9
Truck trying to cross the bridge.

process cannot meet our requirements. We had to adjust the process by having the truck go the extra 10 miles of the long way around.

A process with a Cp ratio equal to 1 is just barely capable of meeting the customers' specifications. Three times out of a thousand, you will not meet your customers' requirements. Would you like to fly an airline that lands safely 997 times out of 1000, or would you like a Cp ratio that is larger than 1? Most people complain loudly about many processes that are just barely capable. They do so without knowing that there may be a process problem.

Let us go back to our moving-day problem. The car my wife was driving was 8 feet wide. The bridge was also 8 feet wide (see Figure 21.10). Did my wife also have to drive the extra 10 miles? She did not, but it was a very tight squeeze through the bridge. In this case the Cp = 1, which told us that the process was capable but there was no room for variation. Any car that is wider will not be able to cross the bridge.

The best situation is a process capability ratio greater than 1 (see Figure 21.11). In this case, the variability in the process is less than the variability allowed by the specifications. It will be an extremely rare instance where your process will produce something defective. This

FIGURE 21.10
Woman trying to cross the bridge.

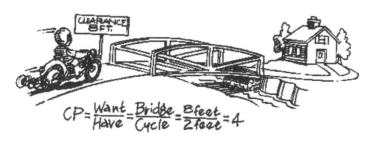

FIGURE 21.11
Motorcycle trying to cross the bridge.

is the type of ratio we depend on in critical components, and when safety or security is a major concern. Processes that have a large Cp ratio should be your target. A way to remember this is "Big is best."

My motorcycle is only 2 feet wide. My Cp is >1, which tells us that my process has a great potential to meet my requirement of going over the bridge. In fact, I can still get over the bridge when unexpected variations occur such as meeting a dog on the bridge. I can zigzag over the bridge with room to spare.

Although the Cp ratio is a powerful measurement of process consistency, it has limitations. It is only a measurement of the potential of the process to meet the specifications. It is not a measure of how much of the product actually falls within the specifications. A process that is centered near either of the specification limits can yield a large amount of bad product even with an excellent Cp ratio. Many customers now require the use of the Cpk ratio to ensure uniformity of products or services that they receive.

I could still run my motorcycle off of the bridge if I was not paying attention, even though I have a lot of extra room on the bridge (see Figure 21.12).

- Cpk ratio

The Cpk ratio equals the Cp ratio when the process is centered on the specifications. When the process is not centered, the Cpk will be smaller than the Cp. The Cpk ratio is also very simple. It uses the USL or the LSL depending on which is closer to center of the curve. See Figure 21.13 for the equation to calculate the Cpk.

The Cpk ratio can help you predict the percentage of nonconforming product. For example, a Cpk ratio of 0.5 is associated with 6.7% out of specification product, whereas a 1.33 Cpk results in 0.003%

FIGURE 21.12
Motorcycle missed the bridge.

$$Cpk = \frac{|\text{ closer specification limit} - \bar{x}\ |}{3\sigma}$$

FIGURE 21.13
Cpk equation.

nonconforming and a 2.0 Cpk is linked to a defect rate of 0.0000001% or 1 defective part per billion. See Table 21.7 for Cpk converted to a nonconforming product.

Many companies think that a 2.0 Cpk is an ideal rate. This may be true if you are producing billions of parts. However, if you are producing a small volume of high-value parts, it may be more cost-effective for you to rigorously sort and inspect your product. You can convert Cpk into the percentage of nonconforming product. Table 21.7 shows these percentages. These estimates apply to a single specification limit of a process that it is normally distributed and in statistical control.

Here is an example of Cpk from a firm that ships packages. The specification is that 80 to 84 packages per day can be delayed. Their mean is 83 packages per day. The standard deviation is 0.5 packages (see Figure 21.14).

TABLE 21.7

Conversion Table

Cpk Converted to Nonconforming Product	
Cpk	**Percent Nonconforming**
0.40	11.5
0.50	6.7
0.60	3.6
0.70	1.8
0.80	0.82
0.90	0.35
1.00	0.14
1.10	0.05
1.20	0.02
1.30	0.005
1.33	0.003
1.50	0.0003
1.60	0.0001
1.67	0.00003
2.00	0.000001%

$$Cp = \frac{84\ pk - 80\ pk}{6(0.5\ pk)} = \frac{4\ pk}{3\ pk} = 1.33$$

$$Cpk = \frac{84\ pk - 83\ pk}{3(0.5\ pk)} = \frac{1\ pk}{1.5\ pk} = 0.67$$

FIGURE 21.14
Late packages.

The process appears capable when Cp is calculated. The Cpk index clarifies that a considerable amount of late packages will result.

Let us go back to our moving-day situation. Remember that you want precision and accuracy. The Cp ratio helped us understand precision. If you remember, our car had a Cp ratio of exactly 1. If the car is going to cross the bridge, it must be very accurate. However, let us look at the motorcycle's Cpk. Cp > 1 tells us that our process (our 2-foot motorcycle) has the potential to meet our requirements (the 8-foot bridge). Our Cpk > 1 also tells us that our process does meet the requirements. Our Cpk tells us that our aim is accurate (see Figure 21.15).

Both indices are used to measure process capability. The Cp index measures the precision of the process. The Cpk index measures both the accuracy of the process. Both indices allow you to

- Concentrate technical know-how and other resources on processes with low Cp and Cpk
- Track improvement in the performance of your process over time
- Improve your aim at your goalpost

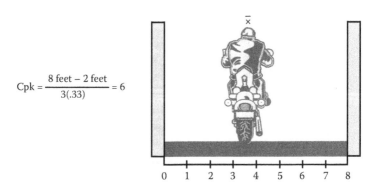

$$Cpk = \frac{8\ feet - 2\ feet}{3(.33)} = 6$$

FIGURE 21.15
Cpk calculations for the motorcycle.

- Using Cp and Cpk

    The larger the value of Cp and Cpk, the better off you are. The larger Cp and Cpk are, the more likely you are able to meet your customers' requirements. Complete a process capability study only after the process is stable. It must consist of variable data and be a normally distributed process. In addition, Cp assumes that the process is centered on the customers' specifications.

    The reason you need to use Cp and Cpk is to prevent defects by taking action on the process rather than having to take action on the output. Sorting and rework are expensive actions to take on your output. It is usually more efficient and effective to fix the process. It is better to quit producing bad products and services rather than to rely on sorting and fixing the customer.

    By using process capability indices, it is possible to

    - Determine which processes are not capable of meeting your customers' needs
    - Identify those processes that are not operating at maximum efficiency
    - Identify the processes that have the least margin of capability
    - Prioritize which processes to improve first
    - Estimate the amount of your output that will be bad
    - Evaluate process performance over extended periods to maintain your best practices

    Today's markets require an ever-increasing proof of quality and consistency of products and services. One tool often used to communicate the amount of variability in what you do is process capability. Several indices of process capability are available. The two most commonly used are the Cp and Cpk indices. Use the one that is most appropriate for you.

Examples of Their Calculation and Interpretation of the Results

Figures 21.16 and 21.17 recap all the indices we have covered (those using SPC as a basis) with examples of their calculation and interpretation of the results. A similar portrayal could be made using the K factor method.

Just for Fun

Laura is getting ready to make some jelly. Unfortunately, she has lost her measuring cups. All she has are two jars. Her counters are full of berries

Condition	#1	#2
USL 46		
MID43		
LSL 40		
Index		
Z	3/3 = 1.0	2/3 = .67
Z_{min}	$\dfrac{46-43}{1}$ or $\dfrac{43-40}{1}$ = 3.0	$\dfrac{46-43}{.67}$ or $\dfrac{43-40}{.67}$ = 4.5
Cp	6/6 = 1.0	6/4 = 1.5
Cpk	3/3 = 1.0	4.5/3 = 1.5
CPU	$\dfrac{46-43}{3}$ = 1.0	$\dfrac{46-43}{2}$ = 1.5
CPL	$\dfrac{43-40}{3}$ = 1.0	$\dfrac{43-40}{2}$ = 1.5
Analysis		
Center condition	Centered	Centered
Inherent capability	Capable	Capable
Operational capability	Minimally capable	Capable
Action	Improve system to add margin of safety	None

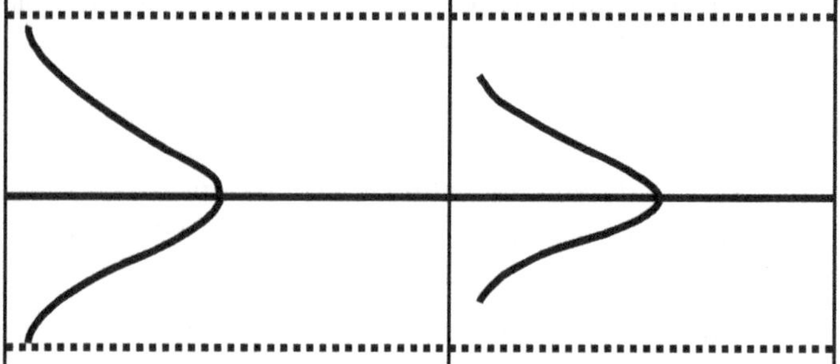

FIGURE 21.16

Example of completed study (page one).

Condition	#3	#4
USL 46	45	42
MID43		
LSL 40		
Index		
Z	$2/3 = .67$	$4/3 = 1.33$
Z_{min}	$\dfrac{46-45}{.67}$ or $\dfrac{45-40}{.67} = 1.5$	$\dfrac{46-42}{1.33}$ or $\dfrac{42-40}{1.33} = 1.5$
Cp	$6/4 = 1.5$	$6/8 = .75$
Cpk	$1.5/3 = .5 \qquad 1.5/3 = .5$	
CPU	$\dfrac{46-45}{2} = .5$	$\dfrac{46-42}{4} = 1.0$
CPL	$\dfrac{45-40}{2} = 2.5 \quad \dfrac{42-40}{4} = .5$	
Analysis		
Center condition	High	Low
Inherent capability	Capable	Incapable
Operational capability	Incapable	Incapable
Action	Center the process	Improve system to

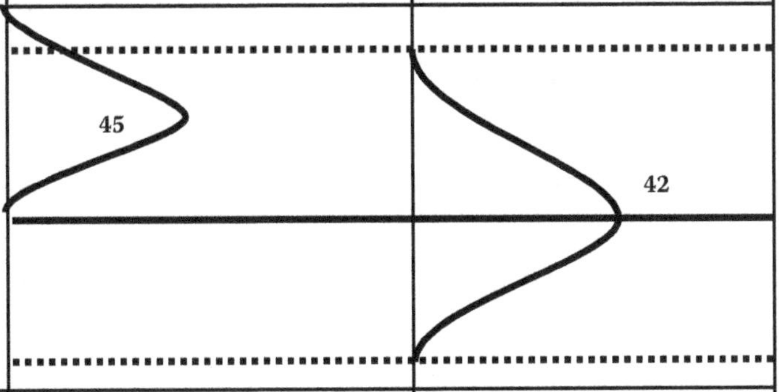

FIGURE 21.17

Example of a completed study (page two).

so she cannot store anything on them. She knows that one jar holds nine cups and the other jar holds four cups. Her recipe calls for six cups of sugar. How can she measure six cups of sugar with only the two jars she has available? The answer:

- Fill nine-cup jar
- Fill four-cup jar from nine-cup jar
- Empty four-cup jar
- Fill four-cup jar from nine-cup jar
- Nine-cup jar now has one-cup of sugar

INTRODUCTION TO EXPERIMENTATION

Where do better ideas and methods come from? Someone has to take a chance and try something new. How did we learn potatoes are good to eat and rocks are not? Someone took a chance and tried to eat a rock, which seemed like a failure. It was a failure to find something good to eat, but it was a success in that learning occurred. Progress requires new learnings. To learn and grow, we must experiment and try new things. It is not easy. Most people do not like to try new things, even just rearranging their furniture.

Experiments are more productive if we plan them. New insights also come from luck. Waiting to trip over new ideas is much less dependable than planning an experiment, carrying it out, carefully checking the results, and then taking action based on the new learnings. This chapter gives a brief introduction to designing experiments. We have found that working with a group of subject matter experts and listing possible new ways of performing a task is a good first step. Running actual experiments to test the new methods identifies and provides hard evidence as to which is the best method. As with other measurement tools, the power lies not in the tool but in the people using the tool. You cannot learn to ride a bike by watching other people ride a bike. You have to experiment and get on one.

Most of what we learn, we learn by screwing it up the first time.

Norman Schwartzkopf

Designing Experiments

How good is good enough? This question has been asked throughout time. There are three ways of answering the question.

1. *Distinguish good from bad.* All you have to do is sort the good stuff from the bad stuff or reduce variation until everything is within the specification. Just as in football, when you kick the ball at the goalpost, you score when the ball goes through the goalposts. Those kicks that go through the goalposts are good and those that do not are bad (see Figure 21.18).
2. *Control variation over time.* When you identify and remove special causes of variation, it puts a process under statistical control (see Figure 21.19).
3. *Drive the process to its target.* This method allows you to focus on what the process should produce and then modify the process to produce the desired result (see Figure 21.20). For example, if a pie is

FIGURE 21.18
Extra point in football.

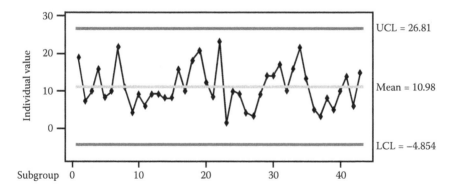

FIGURE 21.19
Total e-Pro transaction by account from September 2002 through March 2003.

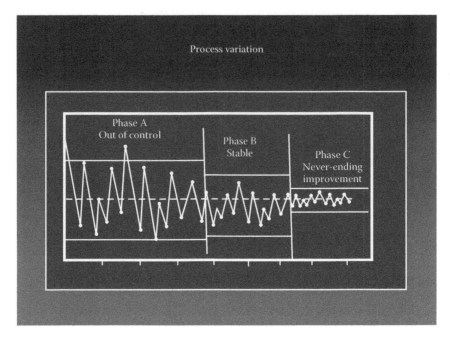

FIGURE 21.20
Reducing variation.

sold as a 24 ounces pie, then it should weigh 24 ounces. This way of looking at it shows that if any pie weighs more than 24 ounces the producer loses money. Pies weighing less than 24 ounces cause the customer to lose value and become upset. The goal is 24 ounces, not 24 ounces plus or minus anything.

No matter which of the three ways you choose to look at your process, experiments can help you get to where you need to go.

When you have decided to improve your process, how are you going to make it better? A well-planned experiment is a tailor-made method to find a better way. Our favorite way uses the following four steps:

1. *Plan your experiment.* When you develop a plan for your experiment, you must begin by determining what outcome is important. You then need to identify what effects the outcome you are looking for.
2. *Do the experiment.* When you are performing the experiment, it is important to have a schedule for running the experiment. You must also observe and record the results.

3. *Study the results.* During this step, you need to look for patterns and relationships between the things you have identified as affecting the outcome.

4. *Take action.* After you have studied the results, you need to develop a new and better way of reaching your desired results.

This four-step experimental method was first developed by Walter Shewhart and fostered by W. Edwards Deming. It is used for designing experiments as well as solving problems.

Your experiment may involve a planned design. If you involve a statistician, they may help you design an experiment differently than you would have in the past. Table 21.8 shows some of the comparisons between how experiments are run in traditional settings and those that use modern methods. The experimental process has several steps and provides new information.

To gain insights into how you might run your process better, you need to first decide what results you would get. Let us go back to the pie example. You may decide that you can make pies more efficiently. You consider many possible ways to make the process better. Some of them are running your oven at a different temperature, increasing or decreasing the amount of time you keep the pies in the oven, or changing the recipe by increasing or decreasing the moisture content.

A well-planned experiment is tailor-made to meet specific objectives and to satisfy practical constraints. After the problem and constraints have been clearly defined, usually a statistician is used to help develop an experimental layout to minimize the required testing effort and maximize the usefulness of the information you get from the experiment.

When designing an experiment to improve a process, you need to talk with those involved. You can do this either in a group or individually. Get them to tell you what they think is important to their customer. Then, get them to describe what they do that effects that result. For example, if the customer likes how moist the final pie is, it is important to identify the variables that might affect the moisture of the pie. Some of those variables might be the temperature of the oven, cooking time, water content before cooking, amount of cornstarch in the pie, thickness of crust, and so forth. It is not unusual to get 12 to 18 variables that you believe may have an effect on the result.

In this case, you think that temperature, cooking time, and the amount of cornstarch in the original pie are probably the key factors that determine

TABLE 21.8

Comparison of Methods Experimentation

Criterion	Traditional	Modern
Basic procedure	Hold everything constant except the factor under investigation.	Plan the experiment to evaluate several factors.
Experimental conditions	Hold material, workers, and machines constant throughout the entire experiment.	Realize the difficulty of holding conditions reasonably constant throughout an entire experiment.
Experimental error	Not measured.	Measured.
Results of evaluation	Evaluate result with a consideration of the amount of experimental error.	Evaluate result by comparing result with the measure of experimental error.
Effect of sequence of experiments	Assume that sequence has no effect.	Guard against by randomization.
Effect of varying both factors simultaneously (interaction)	Not adequately planned into experiment.	Experiment can be planned to include an investigation for interaction between factors.
Validity of results	Results are misleading or erroneous if interaction exists.	Even if interaction exists, a valid evaluation of the results can be made.
Number of measurements	For valid information, more measurements are needed than in the modern approach.	Fewer measurements are needed for useful and valid information.
Definition of problem	Objective of experiment frequently is not defined clearly.	To design an experiment, it is necessary to define the objective in detail.

how moist the final pie is. You then decide to design an experiment to find out if you are correct. You make some pies with a little extra cornstarch and some with a little less cornstarch than normal. Some pies you cook a little more than normal, and some a little less than normal. Some you cook at a little higher temperature than normal, and some at a lower temperature than normal.

In all, you make nine pies. The first pie is your standard pie baked at the normal temperature of 350°F, the normal baking time of 50 minutes, and the normal amount of cornstarch of 1/4 (2/8) of a cup. This pie is perfect. It weighs 184 grams. We call this perfect pie a control sample.

Definition: A control sample is what all of your output will be measured against.

Wow, there are a lot of different pies and combinations to keep track of with nine pies and three different factors that will change. The sequence table in Table 21.9 will make it easier to keep track of what to change and in what order. The first column in this table is what number pie you are going to bake. The factors are the three things you have decided to change: temperature, baking time, and the amount of cornstarch in each pie. You are hoping to find a combination that produces a pie as good as you currently bake that you can produce more efficiently (see Table 21.9).

The nature of conducting experiments can lead you to new results. You need the liberty to learn. You must be willing to try and fail. You may find newer and better ways of doing things or you may not.

When you experiment, the higher temperature will be 375°F. The lower temperature will be 325°F. The extra time will be 55 minutes. The shorter time will be 45 minutes. The cornstarch, which is the third variable, will be 3/8 cup at the high end and 1/8 of a cup at the low end.

Looking at Table 21.9, the first experimental pie will have extra time, extra temperature, and extra cornstarch. Therefore, it will be baked at 475°F for 55 minutes and will have 3/8 of a cup of cornstarch. The second experimental pie will have extra temperature, extra time, and less cornstarch. You will bake it at 375°F for 60 minutes. It will have 1/8 of a cup of cornstarch.

TABLE 21.9

Sequence Table

Factor Trial	X_1	X_2	X_3
1	+	+	+
2	+	+	−
3	+	−	+
4	+	−	−
5	−	+	+
6	−	+	−
7	−	−	+
8	−	−	−

TABLE 21.10

Results of Pie Experiment

Pie #	Temperature	Time (Minutes)	Cornstarch (Cups)	Grams	Result
1	375°F	55	3/8	170	Dry
2	375°F	55	1/8	165	Dusty
3	375°F	45	3/8	172	Crumbly
4	375°F	45	1/8	184	Just right
5	325°F	55	3/8	190	Soft
6	325°F	55	1/8	196	Goopy
7	325°F	45	3/8	199	Runny
8	325°F	45	1/8	205	Soupy

When the pies are done baking, you and your customers try them. The first pie that was cooked at a higher temperature, for a longer time, and with 3/8 cup of cornstarch turns out to be very dry. The last pie cooked for a shorter time, at a lower temperature, with less cornstarch, turns out to be soupy. The pie experiment results given in Table 21.10 show how you baked each of the experimental pies. It also shows the results of each pie as it was tested.

Using a rigorous experimental method enables you to design a well-thought-out process that produces something your customer wants. These results can be achieved through experimentation. The objective is to identify the key customer needs and the variables that make them occur. Be willing to experiment even when your process is running well. In fact, that is the best time to do so!

There are many methods of experimentation. The type of experiment you choose depends a great deal on what you want the experiment to accomplish.

Experiments usually generate an incredible amount of data, often too much to effectively study or interpret. Using random sampling can help you reduce your data to a workable level while still retaining all of the characteristics generated by the experiment.

Behold the turtle: He only makes progress when he sticks his neck out.

James Bryant Conant

Designing experiments is an effective and efficient way to gain insights into how you might run your process better. A well-designed experiment can eliminate the negative effects of two or more variables, while evaluating the interactions between the variables.

Designed experiments are effective because they

- Are laid out to accomplish particular objectives
- Answer specific questions of the person running the experiments
- Attempt to make maximum use of minimum data, time, and dollars
- Are designed to answer specific questions
- Provide the ability to separate the variation due to planned changes from other experimental variation

Designed experiments establish a cause-and-effect relationship between several variables and the outcome being studied because several factors are changed at the same time in a planned way.

Experiments yield critical information about processes in service and manufacturing. There are three main objectives that experiments target:

- Experiments are used to find out whether or not differences exist that can be used to take action.
- Experiments can be used to determine what relationships exist between factors.
- Experiments can measure the effect of varying aspects of a process.

SAMPLING

Sample is a subset of a population that is expected to represent the population. Owing to economic and practical reasons, it is not possible to collect and analyze the data for an entire population. Based on the analysis of a sample, we can draw inference about the population. The sample should be an unbiased representative of the population; that is, all individual members in the population should have an equal chance of being picked as a sample. It is also called *random sampling*.

Sample size depends on the confidence interval and the population size.

If you are manufacturing 100 widgets, you could measure all 100 parts to feel comfortable that they meet requirements. You do not have to wait for a slack season to measure all of the widgets you make. We just did it this way to give you another demonstration that the variation in any measurement will usually follow this bell-shaped normal pattern. An easier way to tell what is happening with your process is to take a sample.

Definition: A sample is a small number or amount of what you produce. It represents your entire output.

Pick out a sample of five widgets from that box of 500. No, no! Not all from the top layer! You need a random sample. Dig around in the box. You cannot accuse us of stacking the numbers.

OK! Now measure these five pieces. To ensure that you get a reasonable representation of widgets, get four more samples of five widgets. Be sure to get the samples in the same way. However, do not take all five samples at one time. Spread the sample, taking time evenly across the entire eight-hour shift. Now, let us see what you got. We recorded the samples in the file chart.

The averages of these five samples vary quite a bit from the actual target of 2.00 inches. We got the average for each sample by adding all of the numbers for the sample and dividing by five. The grand average is 2.001 inches (we added all five averages together and divided by the number of samples [five] to get the grand average). The grand average is also called the mean. We found the range for each sample by subtracting the smallest measurement in the sample from the largest measure in the sample. You will not always come quite so close to the actual average in five samples. However, if you have enough samples, the average of the samples will provide a very good estimate of the actual average of the lot. We recorded the samples in the file chart for transfer to a control chart. You could also record the readings directly on a control chart and save a step.

- Random sampling

  When you use stratification or run experiments, you generate a lot of data. To make the data manageable, a technique called random sampling lets you reduce your data to a manageable level while still retaining all of the characteristics that your experiment originally generated.

  When you collect data, remember:
  - A sample is representative of an entire population only when it is random.
  - Choosing a random sample allows you to make generalizations about the entire population based on information gathered from the sample.
  - Understand your purpose for collecting data. This can help determine which sampling technique is appropriate.

There are four different ways to collect data randomly. Several of them use a random number table. A random number table is a table of numbers that a computer generates with no particular pattern. The purpose of using a random number table is to ensure that no bias is used in choosing your sample. Some pocket calculators and most computers have random number generator keys built in.

Use the sampling method that best meets your needs. The sampling methods are as follows:

- Simple random sampling

    A simple random sample comes when you choose the sample items one at a time from the entire population, so that each member of the population has the same chance of being selected. Use simple random sampling any time the items in a population can be listed without a great deal of effort. Simple random sampling involves using a random number table. There are a few steps to follow when using the simple random sampling method:

    - Number the population.
    - Decide on the sample size.
    - Begin at a predetermined point on the table starting at a different point on the table each time it is used. Move systematically throughout the table.
    - Draw the desired sample size.
    - Choose another number if the number chosen is a duplicate or if the number chosen falls outside of the population size.

- Systematic random sampling

    A systematic random sample is one in which every nth (e.g., every 5th, 12th, 100th) item in the population is chosen for the sample. The population items must form some sort of sequence. Use this technique when it is not possible or convenient to list the entire population. Do not use it where the defects might form a pattern. There are a few steps to follow when using systematic random sampling method:

    - Decide on the sample size.
    - Decide on the interval.
    - Draw the desired sample size by choosing items in the chosen pattern.

    For example, contact every 10th name in the phone book for a marketing survey of what kind of pies they would want to eat and what they would be willing to pay for them.

- Stratified random sampling

    A stratified random sample results from forming subgroups or strata in the population on a basis relevant to the sampling study. Elements with each stratum should be similar. The difference between strata should be great.

    Use the stratified random sampling techniques when logically dividing the population into subgroups. This makes the process less time consuming. There are a few steps to follow when using the stratified random sampling method:
    - Decide on the sample size within each subgroup.
    - Decide on logical subgroups.
    - Draw the desired sample size from each subgroup using a random number table.
- Cluster random sampling

    A cluster random sample results from randomly choosing several subgroups in the population and including all or some members of these subgroups in the sample. Use the cluster random sampling technique when it is more convenient to select items as a group rather than individually. There are a few steps to follow when using the stratified random sampling method:
    - Decide how the population could be grouped for ease of sampling.
    - Decide on sample size.
    - Select at random entire subgroups using a random number table.
    - Analyze all items in, or a random sample of items from, each group selected.

    For example, interview all of the people on just one street in a housing development to study the entire development.

    Decide the frequency of subgroups to be chosen. Subgroups taken very frequently will reflect variations that occur with little external influence, while those taken over a 24-hour period will reflect the effect that operator change, tool wear, etc., have on production. The objective is to detect all the changes the process will normally undergo.

SUMMARY

To attain SPC, the process must first achieve process stability by identifying and eliminating all operational causes of variation. This allows prediction of the performance of the process. The next step is to assess the output to determine process capability. Then, take actions necessary on the system's causes of variation to eliminate all system-caused defects. This drives a never-ending search for improvements to decrease the range of variation while maintaining or monitoring the process to ensure that it stays in a state of statistical control.

The mathematical and probability laws, which are the basis for SPC, need not be thoroughly understood by the individuals using the SPC tools. These tools include such analytical techniques as sampling, frequency distributions, stratification, control charts, and scatter diagrams. Among these, control charts are the workhorse of SPC. Dr. Walter A. Shewhart of Bell Laboratories developed this powerful, but simple, tool to dramatically identify common causes and special causes. While there are several types of control charts, they all have the same basic uses: (a) as a judgment, to give evidence whether a process has been operating in statistical control and to signal the presence of special causes so that corrective action can be taken; (b) as an operation, to maintain statistical control by monitoring current output as a basis for preventative measures.

So, SPC and process improvement is an iterative procedure, repeating the three phases of collecting data, calculating the basis of interpreting the data for statistical control, and when the process is in statistical control, interpreting it for continued process stability. Longitudinal and continuous data from this iterative process enable one to monitor the process, shrink specification limits, and thus constantly improve the process capability.

EXAMPLES

See Figures 21.21 and 21.22.

Kristi's bowling scores—Control chart

Range, Rbar, and natural limits

Product ___Fun___   Process ___Bowling___

Operator _Kristi Smith_   Unit of measure ___Pins___

Date	Month Week	5/1	5/2	5/3	5/4	6/1	6/2	6/3	6/4	7/1	7/2					
Time																
Sample measurements	1	103	98	94	106	104	89	104	95	98	99	\				
	2	99	95	94	100	93	90	99	94	99	95	\				
	3	102	105	93	98	92	92	98	95	100	94		> DATA			
	4	102	106	100	96	90	90	95	95	100	94	/				
	5											/				
Sum Σ		406	404	381	400	379	361	396	379	397	382					
Average		102	101	95	100	95	90	99	95	99	96					
Range		4	11	7	10	14	3	9	1	2	5					

FIGURE 21.21
Typical X-bar R chart.

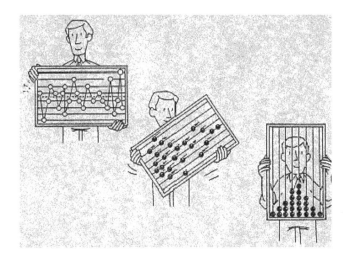

FIGURE 21.22
How a control chart can be converted into a histogram.

SOFTWARE

- MiniTab
- JMP
- Statgraphics
- SAS

SUGGESTED ADDITIONAL READING

American National Standards Institute. Control chart method of controlling quality during production (ASQC Standard B3-1958/ANSI Zi.3-1958, revised 1975). Available from the American Society for Quality Control, 230 West Wells, Milwaukee, WI 53203, for $7.75 ($6.75 to ASQC members).

American National Standards Institute. Guide for quality control and control chart method of analyzing data (ASQC Standards BI-1958 and B2-1958/ANSI Zi. I-1958 and Zi.2-1958, revised 1975). Available from the American Society for Quality Control, 230 West Wells, Milwaukee, WI 53203, for $7.00 ($6.00 to ASQC members).

American Society for Testing and Materials. *Manual on Presentation of Data and Control Chart Analysis (STP-15D)*, 1976. Available from the ASTM, 1916 Race Street, Philadelphia, PA 19103, for $9.75.

22

Tree Diagrams

H. James Harrington

CONTENTS

In QC, we try as far as possible to make our various judgments based on the facts, not on guesswork. Our slogan is Speak with Facts.

Katsuya Hosotani

DEFINITION

Tree diagram breakdown (drilldown) is a technique for breaking complex opportunities and problems down into progressively smaller parts. Start by writing the opportunity statement or problem under investigation down on the left-hand side of a large sheet of paper. Next, write down the points that make up the next level of detail a little to the right of this. These may be factors contributing to the issue or opportunity, information relating to it, or questions raised by it. For each of these points, repeat the process. This process of breaking the issue under investigation into its component part is called drilling down.

USER

This tool can be used by individuals, but its best use is with a group of four to eight people.

OFTEN USED IN THE FOLLOWING PHASES OF THE INNOVATIVE PROCESS

The following are the seven phases of the innovative cycle. An X after the phase name indicates that the tool/methodology is used during that specific phase.

- Creation phase X
- Value proposition phase X
- Resourcing phase
- Documentation phase
- Production phase X
- Sales/delivery phase X
- Performance analysis phase

TOOL ACTIVITY BY PHASE

- Creation, value proposition, production, sales and delivery phases. Note that in solving complex problems, highly creative solutions are needed; tree diagrams are frequently used to divide complex problems into smaller units that can be addressed directly.

HOW TO USE THE TOOL

Tree diagrams are a systematic approach that helps you think about each phase or aspect of solving a problem, reaching a target, or achieving a goal. It is used to break down complex opportunities and problems down into progressively smaller parts, forming a structure that looks like a tree laying on its side, with its trunk breaking out into branches, its branches breaking out into limbs, and its limbs breaking out into leaves.

This approach systematically maps the details of sub- or smaller activities required to complete a project, resolve an issue, or reach a primary goal. This task is accomplished by starting with a key problem or issue and then developing the branches on the tree into different levels of detail. The tree diagram is most often used when a complete understanding of the task is required (i.e., what must be accomplished, how it is to be completed, and the relationships between goals and actions). It can be used at the beginning of the problem-solving process to assist in analyzing a particular issue's root causes before data collection. It can also be used in the final stages of the process to provide detail to a complex implementation plan, allowing for a more manageable approach to the individual elements. Tree diagrams are also used in

- Strategic planning
- Policy development
- Project management
- Change management

The application of this tool is to logically branch out (flowchart) levels of detail on projects, problems, causes, or goals to be achieved (see Figure 22.1).

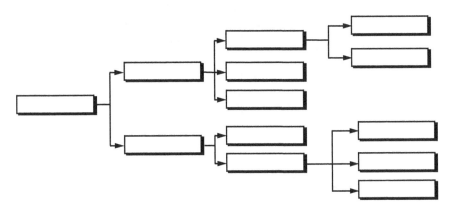

FIGURE 22.1
Typical tree diagram.

There are basically two types of tree diagrams:

- Components-development tree diagram—Typically used in the early stages of the problem-solving process when analyzing a particular problem and trying to establish the root causes before data collection. (This is also known as the *ask why* diagram.) It refers to the components or elements of the work being performed (see Figure 22.2).

    Steps to completing this type of diagram are relatively simple:
    a. State the problem or issue so everyone on the team is in agreement on its meaning. Put that statement in the box on the left of the diagram.
    b. By asking *why*, identify the causes believed to contribute to the problem or issue. Place these causes in a box to the right of the problem or issue. Link them with a line pointing to the cause.
    c. Repeat step b and continue to develop more causes until a key or *root* cause is identified.
- Plan-development tree diagram—Used toward the end of the problem-solving process to provide detail to the individual elements of the implementation plan. (This is also known as the *ask how* diagram.) It helps identify the tasks required to accomplish a goal and the hierarchical relationships between the tasks (see Figure 22.3).

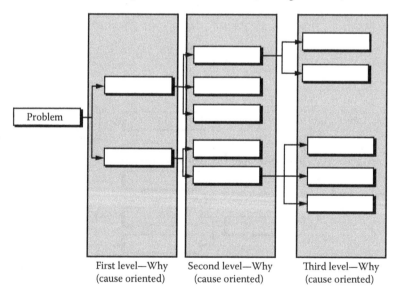

| First level—Why (cause oriented) | Second level—Why (cause oriented) | Third level—Why (cause oriented) |

FIGURE 22.2

Example of a components-development tree diagram.

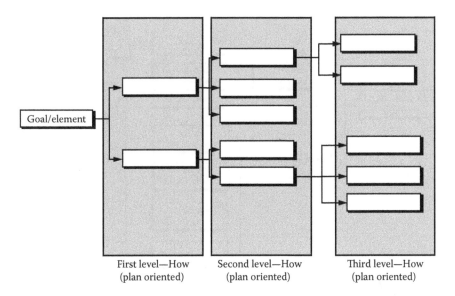

| First level—How | Second level—How | Third level—How |
| (plan oriented) | (plan oriented) | (plan oriented) |

FIGURE 22.3
Example of a plan-development tree diagram.

The plan-development tree diagram is developed much the same way as the component-development diagram. The difference is, here we are trying to provide detail to a particular goal or element of a plan. In step b, instead of asking *why*, you would ask *how* the goal can be achieved. The following are the steps:

a. State the goal or element so everyone on the team is in agreement on its meaning. Put that statement in the box on the left of the diagram.

b. By asking *how*, identify how the goal, task, or element may be achieved. Place this information in a box to the right of the goal/element. Link them with a line pointing to the cause.

c. Repeat step b and continue to develop more detail on the relationship of the task or element, or until the appropriate level of detail has been provided on achieving the goal.

EXAMPLE

Figure 22.4 shows a completed plan-development tree diagram, showing how one element of a total quality management (TQM) improvement plan might look.

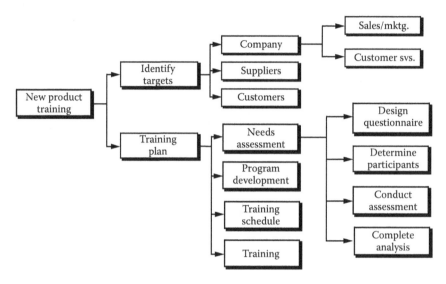

FIGURE 22.4
Example of a semicompleted plan-development tree diagram.

SOFTWARE

Some commercial software available includes but is not limited to

- MindMap: www.novamind.com/
- Smartdraw: www.smartdraw.com/
- QI macros: http://www.qimacros.com

SUGGESTED ADDITIONAL READING

Eiga, T., Futami, R., Miyawama, H., and Nayatani, Y. *The Seven New QC Tools: Practical Applications for Managers.* New York: Quality Resources, 1994.
King, B. *The Seven Management Tools.* Methuen, MA: Goal/QPC, 1989.
Mizuno, S., ed. *Management for Quality Improvement: The 7 New QC Tools.* Portland, OR: Productivity Press, 1988.
Wortman, B. *CSSBB.* West Terre Haute, IN: Quality Console of Indiana, 2001.

23

Value Analysis (Value-Added Analysis)

H. James Harrington

CONTENTS

Less than 50% of resources in most organizations are dedicated to producing the real value-added that a customer would be willing to pay for.

H. James Harrington

DEFINITION

Value analysis is the analysis of a system, process, or activity to determine which parts of it are classified as real-value-added (RVA), business-value-added (BVA), or no-value-added (NVA).

USER

This tool can be used by individuals, but its best use is with a group of four to eight people.

OFTEN USED IN THE FOLLOWING PHASES OF THE INNOVATIVE PROCESS

The following are the seven phases of the innovative cycle. An X after the phase name indicates that the tool/methodology is used during that specific phase.

- Creation phase
- Value proposition phase X
- Resource phase X
- Documentation phase
- Production phase X
- Sales/delivery phase
- Performance analysis phase

TOOL ACTIVITY BY PHASE

- Value proposition and resource phases—During these phases, the costs related to the individual projects are analyzed to determine what percentage of them are RVA and what percentage are BVA or NVA.
- Production phase—During the production phase, it is often necessary to expend effort to reduce costs that is not related to RVA.

HOW TO USE THE TOOL

David Fran, past member of the Economic Development Council of New York City, stated, "Value is the relative cost of providing a necessary function or service at the desired time and place with the essential quality. Value analysis is concerned with identifying the unnecessary costs that do not add essential reliability or quality to the product."

Value-added analysis (VAA) is an analysis of every activity in the business process to determine its contribution to meeting end-customer expectations. The objective of VAA is to optimize RVA activities and minimize or eliminate NVA activities. The organization should ensure

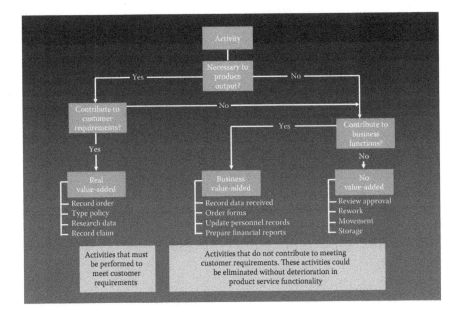

FIGURE 23.1
Value-added assessment.

that every activity within the business process contributes real value to the entire process. Value is defined from the point of view of the external customer's standpoint (see Figure 23.1). There are three classifications of value activities:

- RVA activities—These are the activities that, when viewed by the external customer, are required to provide the output that the customer is expecting. There are many activities performed that are required by the business but are NVA from the external customer's standpoint.
- BVA activities—These are activities that need to be performed to run the organization but which add no value from the external customer's standpoint (e.g., preparing budgets, filling out employee records, updating operating procedures, etc.).
- NVA activities—These are activities that do not contribute to meeting external customer requirements and could be eliminated without degrading the product or service function or the business (e.g., inspecting parts, checking the accuracy of reports, reworking a unit, rewriting a report, etc.). This includes activities classified as bureaucracy activities. There are two kinds of NVA activities:

1. Activities that exist because the process is inadequately designed or the process is not functioning as designed. This includes movement, waiting, setting up for an activity, storing, and doing work over. These activities would be unnecessary to produce the output of the process, but occur because of poor process design. Such activities are often referred to as part of poor-quality cost.
2. Activities not required by the external customer or the process and activities that could be eliminated without affecting the output to the external customer, such as logging in a document.

What every organization has is a huge hidden office made up of BVA and NVA activities. They often account for 80% of the total effort, while RVA activities only account for 20% of the organization's total effort (see Figure 23.2).

Figure 23.3 shows how the evaluation is done. RVA activities contribute directly to producing the output required by the external customer. The process improvement team (PIT) should analyze each activity or task on the flowchart and classify it as an RVA, a BVA, or an NVA activity. (Note: The bureaucracy activities will also be classified as BVA or NVA activities.)

Since this book is not printed in color, you can't visualize the various colors depicted in the Rainbow flowchart. However, we will explain how a Rainbow flowchart would be created in your work environment. You should

FIGURE 23.2
Picture of hidden office.

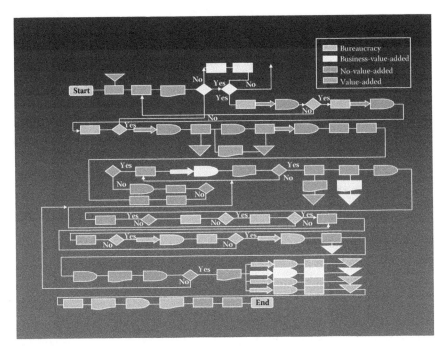

FIGURE 23.3
Rainbow flowchart.

color in the individual different types of activities in the flowchart in order to get in the immediate picture of where the potential problems are occurring. The following is a typical example of how to color in the rainbow flowchart. Use a yellow highlighter to designate each BVA activity on the flowchart. Color in the NVA activities with a red highlighter. You have now turned your flowchart into a rainbow flowchart (see Figure 23.3). Typically, as PIT members go through this phase of the analysis, they are astonished at the small percentage of costs that are RVA activities. Even more alarming is the mismatch of processing time for RVA activities compared to the total processing time. For some business processes, less than 15% of time is spent in RVA activities.

Obviously, this indicates something very wrong, and managers are often disturbed when they learn of these numbers. But there are several explanations:

- As the organization grows, processes break down and are patched for use, thereby making them complex.
- When errors take place, additional controls are put in place to review outputs rather than change the process. Even when the process is corrected, the controls often remain.

- Individuals in the process seldom talk to their customers, and hence do not clearly understand the customer's requirements.
- Too much time is spent on internal maintenance activities (such as coordinating, expediting, record keeping) instead of on doing RVA work.

The individuals doing the analysis should now answer the following questions:

- How can the RVA activities be optimized?
- Can the RVA activities be done at a lower cost with a shorter cycle time?
- How can the NVA activities be eliminated? If they cannot, can they be minimized?
- Why do we need the BVA activities? Can we minimize their cost and cycle time?

The individuals doing the analysis have to be very creative in coming up with solutions and should not be constrained by the current culture, personalities, or environment.

- Rework can be eliminated only by removing the causes of the errors.
- Moving documents and information can be minimized by combining operations, moving people closer together, or automation.
- Waiting time can be minimized by combining operations, balancing work loads, or automation.
- Expediting and troubleshooting can be reduced only by identifying and eliminating the root causes.
- NVA outputs can be eliminated if management agrees.
- Reviews and approval can be eliminated by changes in policies and procedures.

Challenge everything. There is no sacred cow in value analysis activities. Every activity can always be done in a better way. The end result of this analysis is an increase in the proportion of RVA activities, a decrease in the proportion of BVA activities, a minimizing of NVA activities, and a greatly reduced cycle time (see Figure 23.4). This concept is so important that all employees should learn to use it in their daily work. The results will be powerful.

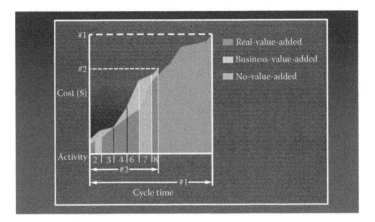

FIGURE 23.4
Cost–cycle time chart before and after applying VAA.

As you analyze each BVA and NVA activity, drive the analysis down to determining the root cause for the activity's existence. Pay particular attention to rework and scrap cycles. We find that the *asking why five times* approach is a simple way of defining the root cause of the activity's existence. Once the root cause is defined, a decision can be made to eliminate, modify, or leave the activity alone. Recently, there have been some breakthroughs in using computer software to help define root causes. We are using a program called Anticipatory Failure Determination produced by I-TRIZ. It is a systematic procedure for identifying the root cause of a failure or other undesirable phenomenon in a process and for making corrections in a timely manner.

Let us take a look at an example of a process we worked on. The object was to eliminate the need for any of the decision blocks. We were working on a decision activity that asked the question, "Does the customer live in a location where the service can be provided?" Answer: Yes, but the location was not in your qualified records.

The following are the results of using the *asking why five times* approach.

- Question 1. Why is the location not qualified?
  - Answer: It is a new area and not in our records.
- Question 2. Why is it not in our records?
  - Answer: Because no one told us they were building in that area.
- Question 3. Why were we not told?
  - Answer: Because we did not look at the records.

- Question 4. Why did we not look at the records?
  - Answer: Because no one was assigned to do it.
- Question 5. Why was no one assigned?
  - Answer: Management did not realize that it would be a problem.

Action: Department 375 will assign John to review the building permits each week and update the records.

This was a very simple example of how this approach drilled down to correcting the problem very effectively. It is amazing how many of these activities can be eliminated through very simple methods, eliminating much of the complexity that we are facing today.

If you figure out what is causing the exceptions, you can streamline the process.

Mark Robertson
EDS Headquarters

Typical Value Analysis Cycle

Everybody is flowcharting their processes but once they are flowcharted, the real challenge begins. The question is, "How can you streamline the process?" The following is one simple approach that has worked for me:

Step 1. Review each block in the flowchart to define which of the following classifications each of the block fits into.
- RVA: These are activities that are directly related to the product or service that will be delivered to the external customer. It is activities that the customer would be willing to pay for (e.g., machining a part, cooking a meal in a restaurant, writing out a sales order, etc.).
- BVA: These are activities that are needed to run the organization but not things that the external customers want you to do for them (e.g., preparing budgets, filling out employee records, updating operating procedures, etc.).
- NVA: These are activities that are not necessary (e.g., inspecting parts, checking the accuracy of report, reworking a unit, rewriting a report etc.). This includes activities already classified as bureaucracy activities.

Now ask yourself if the things you classified as RVA are truly what the customer wants to pay for. The answer in most cases is *yes and no*. Most RVA activities have NVA and/or BVA parts of the activities. For example, let us look at the activity of contacting a customer to give him or her the date when the order will be delivered. It is obvious that this is an activity that the customer wants you to do, but is it all RVA? Probably not. The input to the activity is the order delivery date sent to you over the internal network from production control. Let us look at the tasks that make use of this activity (Table 23.1).

The only part of the activity that is RVA was the part of the conversation with the customer when you provided him or her with the date of the delivery. This was about five seconds of the conversation when you said, "Your order #175 for 50 wheel lugs will be delivered on July 17 in the AM." The rest of the task was BVA or NVA.

You can see that even in the blue-colored activities in the rainbow flowchart, there are significant opportunities to improve. Most of the blue activities have less than 20% of the cost devoted to RVA work.

Be careful—you can have a big problem if you do not consider the NVA cost. (BVA plus NVA equals non-value added.) Table 23.2 shows a product-cost estimate for the normal cost standpoint

TABLE 23.1

Order Delivery Activities

Activity	Type of Value-Added Activity
1. Turn on your computer.	No-value-added
2. Search through the inputs until you find the customer order status.	No-value-added
3. Read the status to be sure it covers all that the customer ordered.	Business-value-added
4. Find your scheduling notebook.	No-value-added
5. Record the data in your notebook.	Business-value-added
6. Look up the customer phone number.	No-value-added
7. Place the call to the customer.	No-value-added
8. Customer calls back.	No-value-added
9. Find your scheduling notebook and find the status of the order.	No-value-added
10. Provide the customer with the date the order will be delivered.	Real-value-added
11. Record in your scheduling notebook that you have completed the task.	Business-value-added
12. Put your scheduling notebook away.	No-value-added

TABLE 23.2

Total Cost versus Functional Cost

Product Cost	Normal Cost	Value-Added Cost	Non-Value-Added Cost	Total Cost
Material	$35.00	$27.50	$7.50	$35.00
Direct labor	$4.50	$3.50	$1.00	$4.50
Storage cost			$6.00	$6.00
Internal moving cost			$2.25	$2.25
Utility cost		$1.80	$0.30	$2.10
Scheduling cost		$0.80		$0.80
Machine cost		$2.00	$1.40	$3.40
Imputed interest		$0.60	$0.60	$1.20
Other mfg cost		$2.08	$1.04	$3.12
Overhead	$10.00	$7.28	$11.59	$18.87
Engineering		$1.20	$0.80	$2.00
Accounting		$1.00	$0.50	$1.50
Mfg. admin		$2.00	$1.60	$3.60
Personnel		$0.75	$1.25	$2.00
Marketing		$4.00	$1.80	$5.80
Sales		$1.20	$2.20	$3.40
Total selling	$8.00	$10.15	$8.15	$18.30
General and admin cost				
Total cost	$57.50	$48.43	$28.24	$76.67
Sales price	$69.99			$69.99
Profit	$12.49			−$6.68

and then what the real costs are when non-value-added costs are considered.

Figure 23.5 is a simple four-block flowchart. The first operation is machining a part, which is RVA; the second operation is recording the time that it takes to machine the part, which is BVA; and the third operation is inspecting to see if it is good or bad, which is NVA.

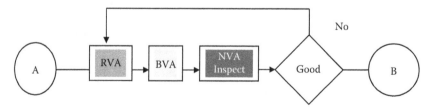

FIGURE 23.5
Process flow diagram.

Step 2. To highlight these differences, we recommend coloring in the RVA activities in green, BVA activities in yellow, and NVA activities in red. This is called a rainbow flowchart. Often, bureaucracy-type, NVA activities are colored in blue to set them apart.

Step 3. For the activities that are classified as NVA, define the ones that are inspection or audit activities.

Step 4. For the inspection and audit activities, define what it costs and how much it delays the process. Next, compare it to the RVA savings that the inspection or audit adds because it keeps defective items from moving on to later higher-cost activities. If the RVA content is less than the cost of doing the audit or inspection, consider eliminating the activity. If the RVA savings is greater than the cost of doing the activity, then continue doing the activity but start a corrective action project to correct the problem to the point that the NVA activity can be eliminated. One word of caution: often, people do a better job because they know that their output will be reviewed or inspected by another person, and if the inspection or audit is dropped, the quality of the work may become unacceptable.

Step 5. For those inspection and audit operations that will remain, collect the defect, error, or deviation data, and plot them using a Pareto diagram (see Table 23.3 and Figure 23.6).

For the top 50% to 60% of the defects, use the 5 *Whys* technique to define the root causes. In some cases, a more sophisticated root cause analysis maybe required.

Step 6. Perform a root cause analysis on the three highest defects: B, E, and F. Table 23.4 shows the results.

For defect B, two root causes were defined (BR1 and BR2), and only one root cause was defined for defects E and F (ER and FR).

TABLE 23.3

Number of Defects by Defect Name

Defect Name	Quality
A	3
B	12
C	2
D	0
E	8
F	10
Misc.	7

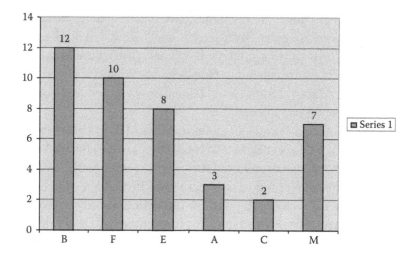

FIGURE 23.6
Pareto diagram of defects.

TABLE 23.4

Root Cause per Defect

Defect	Root Cause
B	BR1
	BR2
E	ER
F	FR

TABLE 23.5

Root Causes and Their Related Corrective Action

Root Cause	Corrective Action
BR1	BR1C1
BR2	BR2C1
	BR2C2
	BR2C3
ER	ERC1
	FRC2
	ERC3
FR	FRC1
	FRC2

Step 7. Now for each of the root causes, a corrective action plan needs to be developed (see Table 23.5).

Step 8. We will now analyze each corrective action to define what impact it would have on the process/organization. Each corrective action is evaluated as a stand-alone item (see Table 23.6).

It is important to understand that a single corrective action can have impact in a number of ways. For example, by reducing variation in a process, process cycle time and cost can be decreased, and quality and customers' satisfaction can be improved. As a result, we analyzed each corrective action in the following ways:

- Dollars saved
- Reduced variation
- Quality improvement
- Cycle time reduction
- Inventory reduction
- Increased customer satisfaction
- Processing time

Step 9. The next activity is to study the impact each of the corrective actions has on the different defects (see Table 23.7).

Step 10. For each corrective action, next evaluate its cost, cycle time, work load, and difficulty to implement (see Table 23.8).

Step 11. By analyzing Table 23.8, define the sequence that the improvements will be implemented. As you define the implementation order, add the improvement to the impact table (see Table 23.9).

In Table 23.9, we recorded the percentage of the defect problems corrected by the corrective action. (Note: Inv. stands for inventory reduction.) Some people prefer to analyze the number of defects that would be eliminated by each corrective action instead of the percent of the problem that will be eliminated. For example, corrective action BR2C1 would reduce the three defects in Table 23.9 problem A by one defect instead of 33%. Both approaches are acceptable.

Each corrective action must assume that the problem has been reduced as defined by the previous corrective action. Example: If the previous corrective action eliminated the burs on a part by using a new tool, a later corrective action could not claim that it had eliminated the burs by changing the materials.

In analyzing the projected impact of improvement for the first three corrective actions, you will note the following results: defects for problem

TABLE 23.6

Impact of Corrective Action

Corrective Action	$ Saved	Reduced Variation	Quality Improvement	Cycle Time	Inventory Reduction	Customer Satisfaction	Processing Time
BR1C1							
BR2C1							
BR2C2							
BR2C3							
ERC1							
ERC2							
ERC3							
FRC1							
FRC2							

TABLE 23.7

Corrective Action Impact on Defects

	Defect Types						
CA	A	B	C	D	E	F	Misc.
BR1C1	+	++	–	0	+	+	0
BR2C1	+	++	+	+	0	0	0
BR2C2	–	++	0	0	+	0	+
BR2C3	0	+	0	0	0	0	0
ERC1	0	×	–	0	++	0	0
ERC2	0	+	0	0	++	+	0
ERC3	0	0	0	0	++	0	0
FRC1	0	+	0	0	+	++	+
FRC2	+	0	0	–	–	+	0

Note: +, Positive; ++, very positive; 0, no impact; –, negative; ×, very negative.

TABLE 23.8

Corrective Action Implementation Analysis

	Cost ($)	Cycle Time (Days)	Difficulty	Work Load (Days)
BR1C1	10,000	30	Medium	5
BR2C1	2,000	10	Low	1
BR2C2	25,000	45	Medium	30
BR2C3	3,000	5	Medium	6
ERC1	1,000	4	Medium	1
ERC2	18,000	18	Medium	10
ERC3	28,000	30	High	60
FRC1	5,000	3	Low	1
FRC2	15,000	10	Medium	8

B are completely eliminated; defects for problem A are reduced by 66%; problem F is solved; and defects for problem E are decreased by 85%.

In the problem E example, after the first three corrective actions were selected and the impacts analyzed, additional improvement was still needed. So the fourth corrective action that was planned for implementation was ERC2. The combination of these four corrective actions was enough to correct the process and five of the potential corrective actions did not need to be implemented.

This approach is a simple and straightforward way of completing a thorough problem–analysis cycle. The first step defines the root causes, and the next steps analyze the potential corrective actions to optimize the positive effects of implementing these corrective actions on the total process.

TABLE 23.9

Implementation Impact

CA	% Defects Reduction							$ Savings	Cycle Time (hr)	Customer Satisfaction	Inv.
	A	B	C	D	E	F	M				
BR2C1	33	40	10	0	0	0	10	25,000	−10	+	+
FRC1	0	30	0	0	15	75	10	30,000	0	+	0
BR1C1	33	30	−10	0	20	25	0	1000	−5	+	+
ERC2	0	0	0	0	50	0	20	1000	0	0	0
Total	66	100	0	0	85	100	40	57,000	−15	3+	2+

Although the 5 Whys is an effective way to drive to the root cause of a problem, there are many other ways to come up with effective corrective action. The following are three of them:

1. Creative questioning
2. Forced idea relationships
3. Lateral type thinking

- Creative questioning

  Alex Osborn originally developed this technique. It has been used extensively in creative activities such as value analysis and visioning. The technique involves identifying an object such as an activity in a process or a physical product, and asking a series of questions about how that object might be changed. Creative questioning may be used either individually or as an aid to brainstorming in a group.
- Forced idea relationships

  Forced idea relationship is a lot like lateral thinking and *morphological* analysis, as it relates to words and idea association to generate new ideas. It usually begins with a brainstorming activity. The group is then asked to force associate at random words that are on the brainstorming list. This forced association then generates creative type thinking.
- Lateral-type thinking

  Lateral thinking is side-wise thinking, that is, to think across categories. Lateral thinking combines word associations with creative cross-categorical thinking. The three primary ways to associate words are
  a. Similarity—this allows you to combine words and ideas based on some common type or category.
  b. Contrast—this allows words and ideas to be associated based on their differences.
  c. Proximity—this allows words and ideas to be associated based on a cause-and-effect relationship or order sequence.

To accomplish this, the group is given key words and is asked to develop other associated words that are combined in a matrix. By examining the matrix, the team comes up with unique associations between key words and the associated words that can lead to innovation.

EXAMPLES

Many of the new light bulbs that now replace incandescent bulbs can be considered examples of value analysis. Other examples were included in the previous section.

SUGGESTED ADDITIONAL READING

Harrington, H.J. *Business Process Improvement*. New York: McGraw-Hill, 1992.

Harrington, H.J. *Streamlined Process Improvement*. New York: McGraw-Hill, 2012.

Harrington, H.J., Gupta, P., and Voehl, F. *The Six Sigma Green Belt Handbook*. Chico, CA: Paton Professional, 2009.

Kaufman, J. *Value Analysis Tear Down*. New York: Industrial Press, 2005.

Park, R.J. *Value Engineering: A Plan for Invention*. Boca Raton, FL: St. Lucie Press, 1998.

Rich, N. and Holweg, M. Value analysis/value engineering. EU Project Report, 2000.

Wortman, B. *CSSBB*. West Terre Haute, IN: Quality Console of Indiana, 2001.

Appendix: Innovation Definitions

The following terms and definitions are all direct quotes from the International Association of Innovation Professionals (IAOIP) "Study Guide to the Basic Certification Exam." The study guide is used to prepare individuals to take the basic examination to be certified as a professional innovator by the IAOIP. These were taken from what the IAOIP is using as the body of knowledge for the innovation professional.

Notes:

1. The terms and definitions that are italicized were not included in the "Study Guide to the Basic Certification Exam" but are included in the methodologies and tools documented technology.
2. In some cases, there is more than one definition for the same tool/methodology. In most of these cases, all the definitions are acceptable and often the additional definitions just help clarify what the methodology/tool is. In some cases, the preferred definition is identified.

TERMS AND DEFINITIONS

5 Whys: A simple but effective method of asking five times why a problem occurred. After each answer, ask why again using the previous response. It is surprising how this may lead to a root cause of the problem, but it does not solve the problem.

5 Whys: A technique to get to the root cause of the problem. It is the practice of asking five times or more why the failure has occurred in order to get to the root cause. Each time an answer is given, you ask why that particular condition occurred. As outlined in the 5 Whys Overview, it is recommended that the 5 Whys be used with Risk Assessment in order to strengthen the use of the tool for innovation and creativity-enhancing purposes.

76 Standard solutions: A collection of problem-solving concepts intended to help innovators develop solutions. A list was developed from referenced works and published in a comparison with the 40 principles to show that those who are familiar with the 40 principles will be able to expand their problem-solving capability. They are grouped into five categories as follows:

1. *Improving the system with no or little change 13 standard solutions*
2. *Improving the system by changing the system 23 standard solutions*
3. *System transitions six standard solutions*
4. *Detection and measurement 17 standard solutions*
5. *Strategies for simplification and improvement 17 standard solutions*

7–14–28 Processes: This is a task-analysis assessment that involves breaking a process down into seven tasks, then breaking it further into 14 tasks, and then another level further into 28 tasks.

40 Inventive principles: The 40 inventive principles that form a core part of the TRIZ (theory of inventive problem solving) methodology invented by G.S. Altshuller. These are 40 tools used to overcome technical contradictions. Each is a generic suggestion for performing an action to, and within, a technical system. For example, principle #1 (segmentation) suggests finding a way to separate two elements of a technical system into many small interconnected elements.

AEIOU frameworks: This is a way to make observations, and stands for activities, environments, interactions, objects, and users. It serves as a series of prompts to remind the observer to record the multiple dimensions of a situation with textured focus on the user and their interactions with their environment.

ARIZ (algorithm for creative problem solving): A procedure to guide the TRIZ student from the statement of the IFR (ideal final result) to a redefinition of the problem to be solved and then to the solutions to the problem.

Absence thinking: Absence thinking involves training the mind to think creatively about what it is thinking and not thinking. When you are thinking about a specific something, you often notice what is not there, you watch what people are not doing, and you make lists of things that you normally forget.

Abstract rules: Abstract rules are those unarticulated, yet essential, guidelines, norms, and traditions that people within a social setting tend to follow.

Abundance and redundancy: Abundance and redundancy is based on belief (not necessarily factual) that if you want a good invention that solves a problem, you need lots of ideas.

Administrative process: This specifies what tasks need to be done and the order in which they should be accomplished, but does not give any, or at least very little, insight as to how those tasks should be realized.

Affinity diagram: Affinity diagram is a technique for organizing a variety of subjective data into categories based on the intuitive relationships among individual pieces of information. It is often used to find commonalties among concerns and ideas. It lets new patterns and relationships between ideas be discovered.

Affordable loss principle: It stipulates that entrepreneurs risk no more than they are willing to lose.

Agile innovation: This is a procedure used to create a streamlined innovation process that involves everyone. If an innovation process already exists, then the procedure can be used to improve the process resulting in a reduction of development time, resources required, costs, delays, and faults.

Analogical thinking and mental simulations: Using past successes applied to similar problems by mental simulations and testing.

Application of technology: These people are intrigued by the inner workings of things. They may be engineers, but even if not, they like to analyze processes, get under the hood, and they like to use technology to solve problems (business or technical).

Architect: Designs (or authorizes others to design) an end-to-end, integrated innovation process, and also promotes organization design for innovation, where each function contributes to innovation capability.

Attribute listing, morphological analysis, and matrix analysis: Attribute listing, morphological analysis, and matrix analysis techniques are good for finding new combinations of products or services. We use attribute listing and morphological analysis to generate new products and services. Matrix analysis focuses on businesses. It is used to generate new approaches, using attributes such as market sectors, customer needs, products, promotional methods, etc.

Attributes-based questions: Questions based on attributes are ones in which you look for a specific attribute of an object or idea.

Balanced breakthrough model: This suggests that successful new products and services are desirable for users, viable from a business perspective, and technologically feasible.

Barrier buster: This helps navigate political landmines and removes organizational obstacles.

Benchmark (BMK): *The standard by which an item can be measured or judged.*

Benchmarking (BMKG): *A systematic way to identify, understand, and creatively evolve superior products, services, designs, equipment, processes, and practices to improve your organization's real performance.*

Benchmarking innovation: A form of contradiction. Doing something completely new—applying an invention in a new way. It means that others are not doing the same thing. Thus, there is nothing to benchmark.

Biomimicry: *Biomimetic or biomimicry is the imitation of the models, systems, and elements of nature for the purpose of solving complex human problems (Wikipedia). It is the transfer of ideas from biology to technology; the design and production of materials, structures, and systems that are modeled on biological entities and processes. The process involves understanding a problem and observational capability together with the capacity to synthesize different observations into a vision for solving a problem.*

Bottom–up planning for innovation: A process where innovations are described in portfolio requirements to meet business objectives.

Brainstorming: *A technique used by a group to quickly generate large lists of ideas, problems, or issues. The emphasis is on quantity of ideas, not quality.*

Brainstorming or operational creativity: Brainstorming combines a relaxed, informal approach to problem solving with lateral thinking. In most cases, brainstorming provides a free and open environment that encourages everyone to participate. While brainstorming can be effective, it is important to approach it with an open mind and a spirit of nonjudgment.

Brainwriting 6-3-5: *An organized brainstorming with writing technique to come up with ideas in the aid of innovation process stimulating creativity.*

Breakthrough, disruptive, new-to-the-world innovation: Paradigm shifts that reframe existing categories. Disruptive innovation drives significant, sustainable growth by creating new consumption occasions and transforming or obsolescing markets.

Bureaucratic process: This occurs where the inputs are defined and a specific routine is performed; however the desired output is obtained only by random chance.

Business case: *A business case captures the reasoning for initiating a project or task. It is most often presented in a well-structure written document, but in some cases may come in the form of a short verbal agreement or presentation.*

Business Innovation Maturity Model (BIMM): This offers a road map to innovation management maturity.

Business model generation canvas: This is a tool that maps what exists. The business model canvas is a strategic management and entrepreneurial tool comprising the building blocks of a business model. The business is expressed visually on a canvas with the articulation of the nine interlocking building blocks in four cluster areas: offering—value proposition; customer—customer segments, customer relationships; infrastructure—distribution channels, key resources, key partnerships, key activities; value—cost structure and revenue model.

Business model innovation: This changes the method by which an organization creates and delivers value to its customers and how, in turn, it will generate revenue (capture value).

Business plan: *A business plan is a formal statement of a set of business goals, the reason they are believed to be obtainable, and the plan for reaching these goals. It also contains background information about the organization or team attempting to reach these goals.*

CO-STAR: *This is specifically designed to focus the creativity of innovators on ideas that matter to customers and have relevance in the market. It is an easy-to-use tool for turning raw ideas into powerful value propositions.*

Capital investment: This is the cost of manufacturing equipment, packaging equipment, and change parts.

Cause-and-effect diagram: *A visual representation of possible causes of a specific problem or condition. The effect is listed on the right-hand side and the causes take the shape of fish bones. This is the reason it is sometimes called a "fishbone diagram" or an "Ishikawa diagram."*

Co-creation innovation: A way to introduce external catalysts, unfamiliar partners, and disruptive thinking into an organization in order to ignite innovation. The term co-creation innovation can be used in two ways: co-development and the delivery of products and services by two or more enterprises; and co-creation of products and services with customers.

Collective effectiveness: In a complex and highly competitive business environment, it is difficult to sustain or support research and development (R&D) and innovation expenses. Networking allows firms access to different external resources like expertise, equipment, and overall know-how that has already been proven with less cost and in a shorter period.

Collective learning: Networking not only helps firms gain access to expensive resources like machinery, laboratory equipment, and technology, but it also facilitates shared learning via experience and good practice sharing events. This brings new insight and ideas for a firms' current and future innovation projects.

Combination methods: A by-product of already applied process, system, product, service wise solutions integrated into a one solution system to produce one end-result that is unique.

Communication of innovation information: Employees vary greatly in their ability to evaluate potentially significant market information and convey qualified information to pertinent receivers in the product development stream.

Comparative analysis: A detailed study/comparison of an organization's product or service to the competitors' comparable product and or service.

Competitive analysis: It consists of a detailed study of an organization's competitor products, services, and methods to identify their strengths and weaknesses.

Competitive shopping (sometimes called mystery shopper): This is the use of an individual or a group of individuals that goes to a competitor's facilities or directly interacts with the competitor's facilities to collect information related to how the competitor's processes, services, or products are interfacing with the external customer. Data is collected related to key external customer impact areas and compared with the way the organization is operating in those areas.

Conceptual clustering: This is the inherent structure of the data (concepts) that drives cluster formation. Since this technique is dependent on language, it is open to interpretation and consensus is required.

Concept tree (see conceptual clustering).

Confirmation bias: *The tendency of people to include or exclude data that does not fit a preconceived position.*

Consumer co-creation: *This means fostering individualized interactions and experience outcomes between a consumer and the producer of the organizational output. This can be done throughout the whole product life cycle. Customers may share their needs and comments, and even help spread the word or create communities in the commercialization phase. This approach provides a one-time limited interaction with consumers. Today, it is possible to enable constant interactions to really transfer knowledge, needs, desires, and trends from the consumer in a more structured way: co-creation.*

Contingency planning: *A process that primarily delivers a risk management strategy for a business to deal with the unexpected events effectively and the strategy for the business recovery to the normal position. The output of this process is called "contingency plan" or "business continuity and recovery plan."*

Contradiction analysis: The process of identifying and modeling contradictory requirements within a system, which, if unresolved, will limit the performance of the system in some manner.

Contradictions: TRIZ defines two kinds of contradictions: physical and technical.

Convergent thinking: Vetting the various ideas to identify the best workable solutions.

Copyrights: Legal protection of original works of artistic authorship.

Core or line extensions, renovation, sustaining close-in innovation: Extends and adds value to an existing line or platform of products via size, flavor, or format. It is incremental improvement to existing products.

Cost–benefit analysis (CBA): *A financial analysis where the cost of providing (producing) a benefit is compared with the expected value of the benefit to the customer, stakeholder, etc.*

Counseling and mentoring: These people love teaching, coaching, and mentoring. They like to guide employees, peers, and even their clients to better performance.

Crazy quilt principle: This is based on the expert entrepreneur's strategy to continuously seek out people who may become valuable contributors to his or her venture.

Create: To make something; to bring something into existence. The difference between creativity and innovation is that the output from innovation has to be a value-added output, while the output from creativity does not have to be value added.

Creative (preferred definition): Using the ability to make or think of new things involving the process by which new ideas, stories, products, etc., are created.

Creative problem solving (CPS): A methodology developed in the 1950s by Osborn and Parnes. The method calls for solving problems in sequential stages with the systematic alternation of divergent and convergent thinking. It can be enhanced by the use of various creative tools and techniques during different stages of the process.

Creative production: These people love beginning projects, making something original, and making something out of nothing. This can include processes or services as well as tangible objects. They are most engaged when inventing unconventional solutions. In an innovation process, these people may thrive on the ideation phase, creating multiple solutions to the identified problems.

Creative thinking: Creative thinking is all about finding fresh and innovative solutions to problems, and identifying opportunities to improve the way that we do things, along with finding and developing new and different ideas. It can be described as a way of looking at problems or situations from a fresh perspective that suggests unorthodox solutions, which may look unsettling at first.

Creativity: Creativity is the mental ability to conceptualize or imagine new, unusual, or unique ideas, to see the new connection between seemingly random or unrelated things.

Cross-industry innovation: This refers to innovations stemming from cross-industry affinities and approaches involving transfers from one industry to another.

Crowdfunding: The collective effort of individuals who network and pool their money, usually via the Internet, to support efforts initiated by other people or organizations.

Crowdsourcing: A term for a varied group of methods that share the attribute that they all depend on some contribution from the crowd. According to Howe, it is the act of a company or institution taking a function once performed by employees and outsourcing it to an undefined (and generally large) network of people in the form of an *open call*.

Culture: Culture is all about how people behave, treat each other, and treat customers.

Culture creator: Ensures the spirit of innovation is understood, celebrated, and aligned with the strategy of the organization.

Customer advocate: Keeps the voice of the customer alive in the hearts, minds, and actions of innovators and teams.

Customer profile: Empathy map is a technique for creating a profile of your customer beyond the simple demographics of age, gender, and income that has been in use for some time.

DVF model (desirable, viable, feasible): Another name for the balanced breakthrough model.

Design innovation: This focuses on the functional dimension of the job to be done, as well as the social and emotional dimensions, which are sometimes more important than functional aspects.

Design for X (DFX): Both a philosophy and methodology that can help organizations change the way that they manage product development and become more competitive. DFX is defined as a knowledge-based approach for designing products to have as many desirable characteristics as possible. The desirable characteristics include quality, reliability, serviceability, safety, user-friendliness, etc. This approach goes beyond the traditional quality aspects of function, features, and appearance of the item.

Design of experiments: This method is a statistically based method that can reduce the number of experiments needed to establish a mathematical relationship between a dependent variable and independent variables in a system.

Directed innovation: Directed innovation is a systematic approach that helps cross-functional teams apply problem-solving methods like brainstorming, TRIZ, creative problem solving, Six Thinking Hats™, Lateral Thinking™, assumption storming, inventing, Question Banking™, and provocation to a specifically defined problem in order to create novel and patentable solutions.

Direction setter: This creates and communicates vision and business strategy in a compelling manner, and ensures innovation priorities are clear.

Disruptive innovation: A process where a product or service takes root initially in simple applications at the bottom of a market and then relentlessly moves upmarket, eventually displacing established competitors.

Divergent thinking: Coming up with many ideas or solutions to a problem.

Diversity trumps ability theorem: This theorem states that a randomly selected collection of problem solvers outperforms a collection of the best individual problem solvers.

Drive to acquire: The drive to acquire tangible goods such as food, clothing, housing, and money, but also intangible goods such as experiences, or events that improve social status.

Drive to bond: The need for common kinship bonding to larger collectives such as organizations, associations, and nations.

Drive to comprehend: People want to be challenged by their jobs, to grow and learn.

Drive to defend: This includes defending your role and accomplishments. Fulfilling the drive to defend leads to feelings of security and confidence.

Edison method: The Edison method consists of five strategies that cover the full spectrum of innovation necessary for success.

Effectuation: Taking action toward unpredictable future states using currently controlled resources and with imperfect knowledge about current circumstances.

Ekvall: Ekvall's model of the creative climate identifies 10 factors that need to be present:
- Idea time
- Challenge
- Freedom
- Idea support
- Conflicts
- Debates
- Playfulness, humor
- Trust, openness
- Dynamism
- Liveliness

Elevator speech: An elevator speech is a clear, brief message or "commercial" about the innovative idea you are in the process of implementing. It communicates what it is, what you are looking for, and how it can benefit a company or organization. It is typically no more than two minutes, or the time it takes people to ride from the top to the bottom of a building in an elevator.

Emergent collaboration: A social network activity where a shared perspective emerges from a group through spontaneous (unplanned) interactions.

Emotional rollercoaster: It is a notion, similar to journey mapping, that identifies areas of high anxiety in a process and, as such, exposes opportunities for new solutions.

Enterprise control: These people love to run projects or teams and control the assets. They enjoy owning a transaction or sale, and tend to ask for as much responsibility as possible in work situations.

Entrepreneur: Someone who exercises initiative by organizing a venture to take benefit of an opportunity and, as the decision maker, decides what, how, and how much of a good or service will be produced. An entrepreneur supplies risk capital as a risk taker, and monitors and controls the business activities. The entrepreneur is usually a sole proprietor, a partner, or the one who owns the majority of shares in an incorporated venture. From the business dictionary.com.

Era-based questions: Era-based questions require that you put yourself in the position of thinking about a question in a different time or place from the one you are currently in.

Experiments: In this context, experiments represent a mixture of surveys and observations in an artificial setting and can be summarized as test procedures.

Ethnography: *Ethnography can be used in many ways, but most significantly in the creation of a new product or service with a clear understanding of the many different ways that a person may accomplish a task based on their own world view. It means observing and recording what people do to solve a problem and not what they say the problems are. It is based on anthropology but used on current human activities. It is based on the belief that what people do can be more reliable than what they say.*

FAST: *An innovative technique to develop a graphical representation showing the logical relationships between the functions of a project, product, process, or service based on the questions "How" and "Why." In this case, it should not be confused with FAST that stands for the Fast Action Solution Team methodology created by H.J. Harrington. The two are very different in application and usage.*

Failure mode effects analysis: A matrix-based method used to investigate potential serious problems in a proposed system prior to final design. It creates a risk priority number that can be used to create a ranking of the biggest risks and then ranks the proposed solution.

Financial management: *Activities and manages of financial programs and operations, including accounting liaison and pay services; budget preparation and execution; program, cost, and economic analysis; and nonappropriated fund oversight. It is held responsible and accountable for the ethical and intelligent use of investors' resources.*

Financial reporting: *Includes the main financial statements (income statement, balance sheet, statement of cash flows, statement of retained earnings, statement of stockholders' equity) plus other financial information such as annual reports, press releases, etc.*

Fishbone diagrams also known as Ishikawa diagrams: A mnemonic diagram that looks like the skeleton of a fish and has words for the major spurs that prompt causes for the problem.

Five dimensions of a service innovation model:
- Organizational
- Product
- Market
- Process
- Input

Flowcharting: *A method of graphically describing an existing or proposed process by using simple symbols, lines, and words to pictorially display the sequence of activities. Flowcharts are used to understand, analyze, and communicate the activities that make up major processes throughout an organization. It can be used to graphically display the movement of product, communications, and knowledge related to anything that takes an input and value to it and produces an output.*

Focus group: *It is made up of a group of individuals that are knowledgeable or would make use of the subject being discussed. The facilitator is used to lead the discussions and record key information related to the discussions.*

Focus group: A focus group is a structured group interview of typically 7 to 10 individuals who are brought together to discuss their views related to a specific business issue. The group is brought together so that the organizer gain information and insight into a specific subject or the reaction to a proposed product.

Force field analysis: *A visual aid for pinpointing and analyzing elements that resist change (restraining forces) or push for change (driving forces). This technique helps drive improvement by developing plans to overcome the restrainers and make maximum use of the driving forces.*

Four dimensions of innovation:
- Technology: technical uncertainty of innovation projects
- Market: targeting of innovations on new or not previously satisfied customer needs
- Organization: the extent of organizational change
- Innovation environment: impact of innovations on the innovation environment

Four-square model: The four-square model is a design process that consists of five steps:
- Problem framing: Identify what problem we intend to solve and outline a general approach for how we will solve it.
- Research: Gather qualitative and quantitative data related to the problem frame.
- Analysis: Unpack and interpret the data, building conceptual models that help explain what we found.
- Synthesis: Generate ideas and recommendations using the conceptual model as a guide.
- Decision making: Conduct evaluative research to determine which concepts or recommendations best fit the desirable, viable, and feasible criteria.

Four-square model for design innovation: Composed of two sets of polar extremes: understand–make and abstract–concrete.

Functional analysis: A standard method of systems engineering that has been adapted into TRIZ. The subject–action–object method is most frequently used now. It is a graphical and primarily qualitative methodology used to focus the problem solver on the functional relationships (good or bad) between system components.

Functional model: A structured representation of the functions (activities, actions, processes, operations) within the modeled system or subject area.

Functional innovation: Involves identifying the functional components of a problem or challenge and then addressing the processes underlying those functions that are in need of improvement. Through this process, overlaps, gaps, discontinuities, and other inefficiencies can be identified.

Futurist: Looks toward the future, scouts new opportunities, helps everyone see their potential. Enables people throughout the organization to discover the emerging trends that most impact their work.

Generic creativity tools: *A set of commonly used tools that are designed to assist individuals and groups to originate new and different thought patterns. They have many common characteristics like thinking positive, not criticizing ideas, thinking out of the box, right brain thinking, etc. Some of the typical tools are benchmarking, brainstorming, six thinking hats, storyboarding, and TRIZ.*

Goal: The end toward which effort is directed: the terminal point in a race. These are always specified in time and magnitude so they are easy to measure. Goals have key ingredients. First, they specifically state the target for the future state and second, they give the time interval in which the future state will be accomplished. These are key input to every strategic plan.

Goal-based questions: These questions pose the end goal without specifying the means or locking you into particular attributes.

Go-to-market investment: This is the cost of slotting fees for distribution, trade spending, advertising dollars (creative development, media spend), promotional programs, and digital/social media.

Gupta's Einsteinian theory of innovation (GETI): This theory states: Thus, every idea must have some energy associated with it that is an outcome of effort and the speed of the thought. Expressed as

- Innovation value = resources × (speed of thought)^2
  where the speed of thought can be described by the following relationship:

- Speed of thought ≡ function (knowledge, play, imagination)

HU diagrams: An effective way of providing a visual picture of the interface between harmful and useful characteristics of a system or process.

Hitchhiking: When a breakthrough occurs, it is a fertile area for innovators. They should hitchhike on the breakthrough to create new applications and improvements that can be inventions.

I-TRIZ: *An abbreviation for ideation TRIZ, which is a restructuring and enhancement of the classical TRIZ methodology based on modern research and practices. It is a guided set of step-by-step questions and instructions that aid teams in approaching, thinking, and dealing with systems targeted for innovation. It provides specific practical team guidance for the following applications:*

- *Solving a nontechnical or business issue*
- *Solving a technical or engineering issue*
- *Finding the root cause(s) of a system issue*

- *Anticipating and preventing possible systems*
- *Predicting and inventing specific innovative products or services customers will want in the future*
- *Patent (invention) evaluation, preparation, and enhancement to either work-around (invent around, design around) an existing (blocking) patent or provide a patent "fence" to prevent possible work-around*

Ideal final result (IFR): This states that in order to improve a system or process, the output of that system must improve (i.e., volume, quantity, quality, etc.), the cost of the system must be reduced, or both. It is an implementation-free description of the situation after the problem has been solved.

Idea priority index: Prioritizes ideas based on the potential cost–benefit analysis, associated risks, and likely time to commercialize the idea.

Idea selection by grouping or tiers: Groups can be helpful in evaluation of tiers like top ideas or worst ideas. Both grouping and tiers are only useful in a batch evaluation process, not a continuous process.

Idea selection by checklist or threshold: An individual idea's list of attributes must match the preset checklist or threshold in order to pass (e.g., be implemented in six months, profit at least $500,000, and require no more than two employees).

Idea selection by personal preference: A manager, director, line employee, or even an expert is used to screen an idea on the basis of his or her own preferences.

Idea selection by point scoring: Uses a scoring sheet to rate a particular idea on its attributes (e.g., an idea that can be implemented in six months gets +5 points, and one that can make more than $\times$ dollars gets +10 points). The points are then added together and the top ideas are ranked by highest total point scores.

Idea selection by priority index (IPI): The IPI prioritizes ideas based on the potential cost–benefit analysis, associated risks, and likely time to commercialize the idea, using the following relationship:
- Annualized potential impact of the idea = ($) $\times$ probability of acceptance
- IPI = annualized cost of idea development ($) $\times$ time to commercialize (years)

Idea selection by voting: Individual(s) can vote openly or in a closed ballot (i.e., blind or peer review). Voting can be weighted or an individual, such as expert, can give multiple votes to a given idea.

Image board, storyboarding, role playing: These are collections of physical manifestations (image collages or product libraries) of the desirable (or undesirable if you are using that as a motivator) to help generate ideas, or to facilitate conversations with users about what they want.

Imaginary brainstorming: *It expands the brainstorming concept past the small group problem-solving tool to an electronic system that presents the problem/opportunity to anyone who is approved to participate in the electronic system. Creative ideas are collected and a smaller group is used to analyze and identify innovative, imaginative concepts.*

Indexing: Providing a tag for a fact, piece of information, or experience, so that you can retrieve it when you want it.

Influence through language and ideas: These people love expressing ideas for the enjoyment of storytelling, negotiating, or persuading. This can be in written or verbal form, or both.

Innovation: *An advancement that transcends a limiting situation within the system under analysis. Another way to describe these limiting situations is to refer to them as contradictory requirements within a system.*

Innovation: *Converting ideas into tangible products, services, or processes. The challenge that every organization faces is how to convert good ideas into profit. That is what the innovation process is all about.*

Innovation (preferred definition): *The process of translating an idea or invention into an intangible product, service, or process that creates value for which the consumer (the entity that uses the output from the idea) is willing to pay for it more than the cost to produce it.*

Innovation benchmarking: Comparing one organization, process, or product to another that is considered a standard.

Innovation blueprint: *A visual map to the future that enables people within an enterprise or community to understand where they are headed and how they can build that future together. The blueprint is not a tool for individual innovators or teams to improve a specific product or service or to create new ones. Rather, the innovation blueprint is a tool for designing an enterprise that innovates extremely effectively on an ongoing basis.*

Innovation culture: A culture that requires continuous learning, practices, and exceptions of risk and failure; holds individuals accountable for an action; and has aggressive timing.

Innovation management: The collection of ideas for new or improved products and services and their development, implementation, and exploitation in the market.

Innovation master plan framework: *The innovation master plan framework consists of five major elements: strategy, portfolio, processes, culture, and infrastructure.*

Innovation metrics: Measurements to validate that the organization innovate. They typically are
- Annual R&D budget as a percentage of annual sales
- Number of patents filed in the past year
- Total R&D headcount or budget as a percentage of sales
- Number of active projects
- Number of ideas submitted by employees
- Percentage of sales from products introduced in the past × year(s)

Innovation process: The innovation process is made up of five phases:
- Phase I: Creation phase
- Phase II: Value proposition phase
- Phase III: Resourcing phase
- Phase IV: Documentation phase
- Phase V: Production phase
- Phase VI: Sales/delivery phase
- Phase VII: Performance analysis phase

Innovative categories: Most service innovations can be categorized into one of the following groups:
- Incremental or radical, based on the degree of new knowledge
- Continuous or discontinuous, depending on its degree of price performance improvements over existing technologies. Sometimes called evolutionary innovation
- Sustaining or disruptive, relative to the performance of the existing products
- Exploitative or evolutionary, innovation in terms of pursuing new knowledge and developing new services for emerging markets

Innovative problem solving: A subset of problem solving in that a solution must resolve a limitation in the system under analysis in order to be an innovative solution.

Innovator: An innovator is an individual who creates a unique idea that is marketable and guides it through the innovative process so that its value to the customer is greater than the resources required to produce.

Insight: A linking or connection between ideas in the mind. The connections matter more than the pieces.

Inspiration: The word inspiration is from the Latin word *inspire*, meaning *to blow into.*

Inspire innovation tools: Tools that stimulate the unique creative powers. Some of them are
- Absence thinking
- Biomimicry
- Concept tree
- Creative thinking
- Ethnography
- HU diagrams
- Imaginary brainstorming
- I-TRIZ
- Mind mapping
- Open innovation
- Storyboarding
- TRIZ

Integrated innovation system: Covers the full end-to-end innovation process, and ensures the practices and tools are aligned and flow easily from one to the other.

***Intellectual property rights:** The expression "intellectual property rights" refers to a number of legal rights that serve to protect various products of the intellect (i.e., "innovations"). These rights, while different from one another, can and do sometimes offer overlapping legal protection.*

Intersection of different sets of knowledge: Networking creates different relationships to be built across knowledge frontiers and opens up the participating organizations to new stimuli and experience.

Intrapreneur: An intrapreneur is an employee of a large corporation who is given freedom and financial support to create new products, services, systems, etc., and does not have to follow the corporation's usual routines or protocols.

Joint risk taking: Since innovation is a highly risky activity, it is very difficult for a single firm to undertake it by itself, and this impedes the development of new technologies. Joint collaboration minimizes the risk for each firm and encourages them to engage in new activities. This is the logic behind many precompetitive consortia collaborations for risky R&D.

Journey map or experience evaluations: A diagram that illustrates the steps your customer(s) go through in engaging with your company, whether it is a product, an online experience, a retail experience, a service, or any combination of these.

Kano analysis: *A pictorial way to look at customer levels of dissatisfaction and satisfaction to define how they relate to the different product characteristics. The Kano method is based on the idea that features can be plotted using axes of fulfillment and delight. This defines areas of must haves, more is better, and delighters. It classifies customer preferences into five categories.*

- *Attractive*
- *One-dimensional*
- *Must-be*
- *Indifferent*
- *Reverse*

Kepner Trego: This method is very useful for processes that were performing well and then developed a problem. It is a good step-by-step method that is based on finding the cause of the problem by asking what changed since the process was working fine.

Key components of successful innovation:
  i. Funding for innovation
  ii. Trained and educated staff
  iii. Collaborative environment
  iv. Key individuals
  v. Corporate infrastructure

Knowledge management (KM): *A strategy that turns an organization's intellectual assets, both recorded information and the talents of its members, into greater productivity, new value, and increased competitiveness. It is the leveraging of collective wisdom to increase responsiveness and innovation.*

Lead users: *Users of a product or service who provide input to the organization related to new products and services because they foresee needs that are still unknown to the marketplace. Lead users innovate, and therefore are considered to be part of the creative consumers' phenomenon, that is, those "customers who adapt, modify, or transform a proprietary offering" (Berthon et al. 2007).*

Leadership metrics: Leadership metrics address the behaviors that senior managers and leaders must exhibit to support a culture of innovation.

Lemonade principle: Based on the old adage that goes, "If life throws you lemons, make lemonade." In other words, make the best of the unexpected.

Link between climate and organizational innovation: Nine areas need to be evaluated to determine this linkage:

- Challenge, motivation
- Freedom
- Trust
- Idea time
- Play and humor
- Conflicts
- Idea support
- Debates
- Risk taking

Live-ins, shadowing, and immersion labs: They are designed to resemble the retail or home environment and gather extensive information about product purchase or use. These laboratories are used to both test the known, launch new product, and to observe user behavior.

Lotus blossom: This technique is based on the use of analytical capacities and helps generate a great number of ideas that will possibly provide the best solution to the problem to be addressed by the management group. It uses a six-step process.

Managing people and relationships: Unlike counseling and mentoring people, these people live to manage others on a day-to-day basis.

Marketing research: Can be defined as the systematic and objective identification, collection, analysis dissemination, and use of information that is undertaken to improve decision making related to products and services that are provided to external customers.

Market research tools: The following are typical marketing research tools:

- Analysis of customer complaints
- Brainstorming
- Contextual inquiry, empathic design
- Cross-industry innovation
- Crowdsourcing
- In-depth interview
- Lead user technique
- Listening-in technique
- Netnography

- Outcome-driven innovation
- Quality function deployment
- Sequence-oriented problem identification, sequential incident technique
- Tracking, panel
- Analytic hierarchy process
- Category appraisal
- Concept test, virtual concept test
- Conjoint analysis
- Store and market test
- Free elicitation
- Information acceleration
- Information pump
- Kelly repertory grid
- Laddering
- Perceptual mapping
- Product test, product clinic
- Virtual stock market, securities trading of concepts
  i. Zaltman metaphor elicitation technique
  ii. Customer idealized design
  iii. Co-development
  iv. Expert Delphi discussion
  v. Focus group
  vi. Future workshop
  vii. Toolkit

Matrix diagram (decision matrix): *A systematic way of selecting from larger lists of alternatives. They can be used to select a problem from a list of potential problems, select primary root causes from a larger list, or to select a solution from a list of alternatives.*

Medici effect: The book by this name describes the intersection of significantly different ideas that can produce cross-pollination of fields and create more breakthroughs.

Mentor: Coaches and guides innovation champions and teams.

Methodology merger: Each methodology brings with it certain strengths and weaknesses that serve to fulfill specific steps and activities represented on the problem-solving pathway. When combined together and properly utilized, these methodologies create a very effective and useful outcome.

Mind mapping: *An innovation tool and method that starts with a main idea or goal in the middle, and then flows or diagrams ideas out from this one main subject. By using mind maps, you can quickly identify and understand the structure of a subject. You can see the way that pieces of information fit together in a format that your mind finds easy to recall and quick to review. They are also called spider diagrams.*

Mini problem: One that is solved without introducing new elements. We have to understand resources, since the emphasis is on solving the problem without introducing anything new to the system.

Moccasins/walking in the customer's shoes: The moccasins approach is more often called walking in the customer's shoes. This activity allows members of the organization to directly participate in the process that the potential customer is subjected to by physically playing the role of the customer.

Myers–Briggs (MBTI): This is a survey-style measurement instrument used in determining an individual's social style preference.

Network-centric approach: The network-centric approach is taught in colleges and based on collaborative brainstorming. The concept is that more minds are better than one at a given time.

Networker: Works across organizational boundaries to engage stakeholders, promotes connections across boundaries, and secures widespread support.

Nominal group technique: *A technique for prioritizing a list of problems, ideas, or issues that gives everyone in the group or team equal voice in the priority setting process.*

Nonalgorithmic interactions: Actions with cognitive and physical materials of a project whose results you cannot predict for certain; those results you do not know.

Nonprobability sampling techniques: Use samples drawn according to specific and considered characteristics and are therefore based on the researcher's subjective judgment.

Nonprofit: An organization specifically formed to provide a service or product on a not-for-profit basis as determined by applicable law.

NSD: An abbreviation for *new service development.*

Observation: In this context, observation means the recording of behavioral patterns of people, objects, and events in order to obtain information.

Online collaboration: Convening an online brainstorming or idea generation session so members can participate remotely, instead of organizing a group in a room together.

Online management platforms: These are used to foster innovation and enable large groups of people to innovate together—across geographies and time zones. Users can post ideas and value propositions online and can collaborate with others to make these stronger. The community can rate and rank ideas or value propositions, post comments and recommendations, link to resources, build on each other's ideas, and support each other to improve each other's innovations.

Open innovation: The use and application of collective intelligence to produce a creative solution to a challenging problem, as well as to organize large amounts of data and information. The term refers to the use of both inflows and outflows of knowledge to improve internal innovation and expand the markets for external exploitation of innovation. The central idea behind open innovation is that, in a world of widely distributed knowledge, companies cannot afford to rely entirely on their own research, but should instead buy or license processes or inventions (i.e., patents) from other companies.

Opportunity-driven model: Opportunity-driven model is more representative of street-smart individuals who take an idea at the right time and the right place, devise a solution, know how to market it, and capitalize on their breakthrough. They also appear to be lucky.

Organizational capability metrics: Organizational capability metrics focus on the infrastructure and process of innovation. Capability measures provide focus for initiatives geared toward building repeatable and sustainable approaches to invention and reinvention.

Organizational change management (OCM): A comprehensive set of structured procedures for the decision-making, planning, executing, and evaluation activities. It is designed to minimize the resistance and cycle time to implementing a change.

Organizational effectiveness measurements: The following is a typical way of measuring the organization's innovation effectiveness. Typically, it is measured in four key areas of management processes: product, sales, internal services, and sales and marketing. Each area is typically evaluated in a combination of the following:

- Committed leadership
- Clear strategy
- Market insights
- Creative people
- Innovative culture
- Competitive technologies
- Effective processes
- Supportive infrastructure
- Managed projects

Organization internal boundaries: Employee silos often isolate chains of command and communication, which can impede the progress of a valuable idea through product development.

Osborn method: Original brainstorming method developed by Alex F. Osborn by primarily requiring solicitation of unevaluated ideas (divergent thinking), followed by convergent organization and evaluation.

Outcome-driven innovation (ODI): *Built around the theory that people buy products and services to complete tasks or jobs they value. As people complete these jobs, they have certain measurable outcomes that they are attempting to achieve. It links a company's value creation activities to customer-defined metrics. Included in this method is the opportunity algorithm, which helps designers determine the needs that satisfied customers has. This help determine which features are most important to work on. Most important is this tool's intention of trying to find unmet needs that may lead to new and innovative products/services.*

PESTEL frameworks: The PESTEL framework focuses on the macroeconomic factors that influence a business. These factors are

- Political factors
- Economic factors
- Social factors
- Technological factors
- Environmental factors

Patents: A government-granted right that literally and strictly permits the patent owner to prevent others from practicing the claimed invention.

Performance engine project: A project that seeks to improve a current level of performance and not to create a new value proposition.

Permeability to innovation idea sources: Information and idea seeking differs greatly among companies.

Physical contradictions: Situations where one requirement has contradictory, opposite values to another.

Pilot in the plane principle: On the basis of the concept of *control*, using effectual logic, and is referred to as *nonpredictive control*. Expert entrepreneurs believe they can determine their individual futures best by applying effectual logic to the resources they currently control.

Pipeline model: Pipeline model, as driven by chance or innate genius, is a somewhat common perception of the innovation process. Inventors who work in research drive the pipeline model and development environment on a specific topic, explore new ideas, and develop new products and services.

Plan–do–check–act (PDCA): *A structured approach for the improvement of services, products, and/or processes. It is also sometimes referred to as plan–do–check–adjust. Another version of this PDCA cycle is OPDCA. The added "O" stands for observation or, as some versions say, "Grasp the current condition."*

Platform: A consumer need–based opportunity that inspires multiple innovation ideas with a sustainable competitive advantage to drive growth.

Portfolio management: The ongoing management of innovation to ensure delivery against stated goals and innovation strategy.

Post-Fordist: Companies after the Henry Ford efficient production era where managers wielded inordinate responsibility for profit and loss, and the new postmodern leaders of the global economy, who are responsible for developing talented teams.

Potential investor presentation: *A short PowerPoint presentation designed to convince an individual or group to invest their money in an organization or a potential project. It can be a presentation to an individual or group not part of the organization, or the management of the organization that the presenter is presently employed by. It is usually part of a short meeting that usually lasts no more than 1 hour.*

Practices: To look at all the inputs that we have available for selection and all the available operations or routines that we can perform on those inputs, then to select those inputs and operations that will give us our desired results.

Primary data: Data collected from the field or expected customer.

Primary sources: Gathered directly from the source; for instance, if new customer opinions were required to justify a new product, then customer interviews, focus groups, or surveys would suffice.

Principles of invention: A set of 40 principles from a variety of fields such as software, health care, electronics, mechanics, ergonomics, finance, marketing, etc., used to solve problems.

Proactive personal creativity: Proactive strategies to be especially effective in increasing the originality and effectiveness of personal creativity:

- Self-trust
- Open up
- Clean and organize
- Make mistakes
- Get angry
- Get enthusiastic
- Listen to hunches
- Subtract instead of adding
- Physical motion
- Question the questions
- Pump up the volume
- Read, read, read

Probability sampling techniques: Use samples randomly drawn from the whole population.

Probe-and-learn strategy: Where nonworking prototypes are developed in rapid succession, tested with potential customers, and feedback is sought on each prototype.

Problem detection and affinity diagrams: Focus groups, mall intercepts, or mail and phone surveys that ask customers what problems they have. They are all forms of problem detection. The responses are grouped according to commonality (affinity diagrams) to strengthen the validity of the response. Developing the correct queries and interpreting the responses are critical to the usefulness of the method.

Problem solving: Generating a workable solution.

Process: A series of interrelated activities or tasks that take an input and produces an output.

Process phases:
- Phase 1: Opportunity identification
- Phase 2: Idea generation
- Phase 3: Concept evaluation
- Phase 4: Acquiring resources
- Phase 5: Development
- Phase 6: Producing the product
- Phase 7: Launch
- Phase 8: Sales and marketing
- Phase 9: Evaluation of results

Process innovation: Innovation of internal processes. New or improved delivery methods may occur in all aspects of the supply chain in an organization.

Process redesign: *A methodology used to streamline a current process with the objective of reducing cost and cycle time by 30% to 60% while improving output quality from 20% to 200%.*

Process reengineering: *A methodology used to radically change the way a process is designed by developing an aggressive vision of how it should perform and using a group of enablers to prepare a new process design that is not hampered by the present processes paradigms. Use when a 60% to 80% reduction in cost or cycle time is required. Process reengineering is sometimes referred to as new process design or process innovation.*

Product innovation: A multidisciplinary process usually involving many different functions within an organization and, in large organizations, often in coordination across continents.

Project management: *The application of knowledge, skills, tools, and techniques to project activities in order to meet or exceed stakeholders' needs and expectations from a project. (Source: PMBOK Guide.)*

Proof of concept (POC): *A demonstration, the purpose of which is to verify that certain concepts or theories have the potential for real-world application. A proof of concept is a test of an idea made by building a prototype of the application. It is an innovative, scaled-down version of the system you intend to develop. The proof of concept provides evidence that demonstrates that a business model, product, service, system, or idea is feasible and will function as intended.*

Pyramiding: A search technique in which the searcher simply asks an individual (the starting point) to identify one or more others who he or she thinks has higher levels of expertise.

Qualitative research (survey): Represents an unstructured, exploratory research methodology that makes use of psychological methods and relies on small samples, which are mostly not representative.

Qualitative research: Gathered data is transcribed, and single cases are analyzed and compared in order to find similarities and differences to gain deeper insights into the subject of interest. In *quantitative research*, the data preparation step contains the editing, coding, and transcribing of collected data.

Quality function deployment (QFD), also known as the house of quality: This creates a matrix that looks like a house that can mediate the specifications of a product or process. There are subsequent derivative houses that further mediate downstream implementation issues.

Quantitative analysis: These people love to use data and numbers to figure out business solutions. They may be in classic quantitative data jobs, but may also like building computer models to solve other types of business problems. These people can fall into two camps: (i) descriptive and (ii) prescriptive.

Quantitative research (survey): Can be seen as a structured research methodology based on large samples. The main objective in quantitative research is to quantify the data and generalize the results from the sample, using statistical analysis methods.

Quickscore creativity test: A three-minute test that helps assess and develop business creativity skills.

ROI metrics: ROI metrics address two measures: resource investments and financial returns. ROI metrics give innovation management fiscal discipline and help justify and recognize the value of strategic initiatives, programs, and the overall investment in innovation.

Radical innovation: A high level of activity in all four dimensions, while incremental innovations (low degree of novelty) are only weakly to moderately developed in the four dimensions.

Ranking or forced ranking: Ideas are ranked (#1, 2, 3, etc.)—this makes the group consider minor differences in ideas and their characteristics. For forced ranking, there can only be a single #1 idea, a single #2 idea, and so on.

Rating scales: An idea is rated on a number of preset scales (e.g., an idea can be rated on a 1 to 10 on implementation time; any idea that reaches a 9 or 10 is automatically accepted).

Reverse engineering: *This is a process where organizations buy competitive products to better understand how the competitor is packaging, delivering, and selling their product. Once the product is delivered, it is tested, disassembled, and analyzed to determine its performance; how it is assembled; and to estimate its reliability. It is also used to provide the organization with information about the suppliers that the competitors are using.*

Robust design: *Robust design is more than a tool; it is complete methodology that can be used in the design of systems (products or processes) to ensure that they perform consistently in the hands of the customer. It comprises a process and tool kit that allows the designer to assess the impact of variation that the system is likely to experience in use, and if necessary redesign the system if it is found to be sensitive.*

Role model: Provides a living example of innovation through attention and language, as well as through personal choices and actions. Key stakeholders often test the leader's words, to see if these are real. For innovation to move forward, the leader must pass these inevitable tests—to show that, yes, he or she is absolutely committed to innovation as essential to success.

Root cause analysis (RCA): A graphical and textual technique used to understand complex systems and the dependent and independent fundamental contributors, or root causes, of the issue or problem under analysis. This is a technical process in that it provides specific direction as to how to execute the method.

Rote practice: Those activities where it looks like people are engaged in finding the right routines and inputs to obtain the desired result, but are just going through the motions.

S-curve: *A mathematical model also known as the logistic curve, which describes the growth of one variable in terms of another variable over time. S-curves are found in many fields of innovation, from biology and physics to business and technology.*

SCAMPER: *A tool that helps people to generate ideas for new products and services by encouraging them to think about how you could improve existing ones by using each of the six words that SCAMPER stands for and applying it to the new product or service in order* *to generate additional new ideas. SCAMPER is a mnemonic that stands for*

- Substitute
- Combine
- Adapt
- Modify
- Put to another use
- Eliminate
- Reverse

SIPOC: An acronym for *supplier, input, process, output,* and *customer* model.

Scarcity of innovation opportunities: Markets have matured into commoditized exchanges.

Scenario analysis: A process of analyzing possible future events by considering alternative possible outcomes (sometimes called "alternative worlds"). Thus, the scenario analysis, which is a main method of projections, does not try to show one exact picture of the future. Instead, scenario analysis is used as a decision-making tool in the strategic planning process in order to provide flexibility for long-term outcomes.

Scientific method: The classical method that uses a hypothesis based on initial observations and validation through testing and revision if needed.

Secondary data: Data collected through in-house (desk research).

Secondary data sources: Involve evidence gathered from someone other than the primary source of the information. Most media outlets, magazines, books, articles, trade journals, market research reports, or publisher-based information are considered secondary sources of evidence.

Service innovation: Not substantially different than product innovation in that the goal is to satisfy customers' jobs-to-be-done, wow and retain customers, and ultimately optimize profit.

Seven key barriers to personal creativity: Seven key barriers to personal creativity:

- Perceived definitions of creativity
- Presumed uses for creativity
- Overdependence on knowledge
- Experiences and expertise
- Habits

- Personal and professional relationship networks
- Fear of failure

Simulation: *As used in innovation, simulation is the representation of the behavior or characteristics of one system through the use of another system, especially a computer program designed for the purpose. As such, it is both a strategy and a category of tools—and is often coupled with CAI (computer-aided innovation), which is an emerging simulation domain in the array of computer-aided technologies. CAI has been growing as a response to greater industry innovation demands for more reliability in new products.*

Six Sigma: A method designed for the reduction of variation in processes. The general steps used within the DMAIC (define, measure, analyze, improve, and control) and DMADV (define, measure, analyze, develop, and verify) methodologies are mostly administrative in nature. Combining Six Sigma with other tool sets pushes the process strongly toward the technical end of the scale.

Six Thinking Hats: *It is used to look at decisions from a number of important perspectives. This forces you to move outside your habitual thinking style, and helps you to get a more rounded view of a situation. The thinking is that if you look at a problem with the "Six Thinking Hats" technique, then you will solve it using any and all approaches. Your decisions and plans will mix ambition, skill in execution, public sensitivity, creativity, and good contingency planning.*

Social business: The practice of using social technologies to transform business.

Social innovation: Social innovation relates to creative ideas designed to address societal challenges—cultural, economic, and environmental issues—that are no longer simply a local or national problem but affect the well-being of the planet's inhabitants and ultimately, corporate profits and sustainability.

Social media: Refers to using social technologies as media in order to influence large audiences.

Social networks: *Networks of friends, colleagues, and other personal contacts: strong social networks can encourage healthy behaviors. They are often an online community of people with a common interest who use a website or other technologies to communicate with each other and share information, resources, etc. A business-oriented social network is a website or online service that facilitates this communication.*

Spontaneous order: A term that Hayek uses to describe what he calls the open society. It is created by unleashing human creativity generally in a way not planned by anyone, and, importantly, could not have been.

Stage gate process: First introduced by R.G. Cooper in 1986 in his book *Winning at New Products*.

Stakeholder: *A "stakeholder" of an organization or enterprise is someone who potentially or really influences that organization, and who is likely to be affected by that organization's activities and processes. Or, even more significantly,* **perceives** *that they will be affected (usually negatively).*

Statistical analysis: *A collection, examination, summarization, manipulation, and interpretation of quantitative data to discover its underlying causes, patterns, relationships, and trends.*

Storyboarding: *Physically structuring the output into a logical arrangement. The ideas, observations, or solutions may be grouped visually according to shared characteristics, dependencies on one another, or similar means. These groupings show relationships between ideas and provide a starting point for action plans and implementation sequences.*

Substantial platform, transformational, adjacencies innovation: Innovations that deliver a unique or new benefit or usage occasion, within an existing or adjacent category.

Synectics: *It combines a structured approach to creativity with the freewheeling problem-solving approach used in techniques like brainstorming. It is a useful technique when simpler creativity techniques like SCAMPER, brainstorming, and random input have failed to generate useful ideas. It uses many different triggers and stimuli to jolt people out of established mind-sets and into more creative ways of thinking.*

Systematic innovation stages: Systematic innovation can be viewed as occurring in stages:
- Concept stage
- Feasibility stage
- Development stage
- Execution stage, preparation for production
- Production stage
- Sustainability stage

Systematic innovation tools:

- Analogical thinking and mental simulations
- Theory of inventive problem solving (TRIZ)
- Scientific method
- Edison method
- Brainstorming
  i. Osborn method
  ii. Six Thinking Hats
  iii. Problem detection and affinity diagrams
  iv. Explore unusual results
  v. Ethnography
  vi. Function analysis and fast diagrams
  vii. Kano method
  viii. Abundance and redundancy
  ix. Hitchhiking
  x. Kepner Trego
  xi. Quality function deployment (QFD), also known as the house of quality
  xii. Design of experiments
  xiii. Failure mode effects analysis
  xiv. Fishbone diagrams, also known as Ishikawa diagrams
  xv. Five whys
  xvi. Medici effect
  xvii. Technology mapping and recombination
- Trial and error

System operator (also called *nine windows* or *multiscreen* method): A visual technique that is used frequently in the initial stages of TRIZ as part of problem definition.

System operator: The construction of a 3 × 3 matrix, with the rows labeled as the system, subsystem, super system; and the columns labeled past, present, and future.

Systems engineering: These methods are more technical than administrative processes, as they are fairly specific as to how to create and utilize the various systems engineering models. However, these methods guide the problem solver understanding that full system analysis is necessary in creating truly effective solutions. Therefore, these methods may be more administrative in nature than technical.

Systems engineering, system analysis: A technique to ensure that full-system effects, impacts, benefits, and responses are understood when looking at changes or problems within a system.

Systems thinking: An approach to problem solving, by viewing "problems" as parts of an overall system, rather than reacting to specific parts, outcomes, or events, and potentially contributing to further development of unintended consequences.

TEDOC methodology: TEDOC stands for target, explore, develop, optimize, and commercialize data to solve problems creatively. As such, TRIZ brings repeatability, predictability, and reliability to the problem-solving process with its structured and algorithmic approach.

TRIZ (pronounced "treesz"): A Russian acronym for "Teoriya Resheniya Izobretatelskikh Zadatch," the theory of inventive problem solving, originated by Genrich Altshuller in 1946. It is a broad title representing methodologies, toolsets, knowledge bases, and model-based technologies for generating innovative ideas and solutions. It aims to create an algorithm to the innovation of new systems and the refinement of existing systems, and is based on the study of patents of technological evolution of systems, scientific theory, organizations, and works of art.

Technical contradictions: The classical engineering and management trade-offs. The desired state cannot be reached because something else in the system prevents it. The TRIZ patent research classified 39 features for technical contradictions.

Technical process: Specifies not only what needs to be done, and in what order, but also provides specific details of *how* to execute the various tasks.

Technically focused brainstorming: The use of standard brainstorming methods bounded by certain acceptable solution concept conditions and guided by the attainment of an *ideal solution*.

Technically focused brainstorming: This methodology guides the generation of solution concepts by ensuring that those solution concepts support the resolution of contradictory requirements of the system under analysis, and renders that system to be of higher value than it was before the solution was applied.

Technology mapping and recombination: A matrix-based method that lists the various technologies that can perform a function and then examines combinations that have not been tried to see if there is enhanced performance or features.

Theory development and conceptual thinking: These people love thinking and talking about abstract ideas. They love the *why* of strategy more than the *how*. They may enjoy business models that explain the reasons behind the competitive position of a business.

Things are not innovation: The following maintenance or change management activities are *not* types of innovation:

- Cost savings
- Ingredient or product changes
- Regulatory change
- Label change

Thinking innovatively: Thinking innovatively is soliciting ideas from everyone, which is a challenge. There is a need for training people in asking questions, thinking of ideas, and articulating their ideas in words or graphics.

Thrashing: A term used to describe ineffective human workgroup activity, effort lost in unproductive work.

Time-of-day map: This tool focuses the participants not on a task or an experience, but rather what happens or does not happen in two- to four-hour chunks in a person's day and what opportunities may appear.

Top–down planning for innovation: Generally, a revenue goal-driven process that is usually set from the top by the senior leadership team. It is usually a dollar revenue goal or a percentage of revenue target from innovation.

Trademarks: Words or logos that are used by someone to identify their products or services, and distinguish them from the words or logos of others.

Trade secrets: Essentially refers to the legal protection often granted to confidential information having at least potential competitive value.

Tree diagram breakdown (drilldown): *A technique for breaking complex opportunities and problems down into progressively smaller parts. Start by writing the opportunity statement or problem under investigation down on the left-hand side of a large sheet of paper. Next, write down the points that make up the next level of detail a little to the right of this. These may be factors contributing to the issue or opportunity, information relating to it, or questions raised by it. For each of these points, repeat the process. This process of breaking the issue under investigation into its component part is called drilling down.*

Trends: Those dimensions on which lead users are far ahead of the mass market.

Trial and error: Attempts at successful solutions to a problem with little benefit from failed attempts.

Value analysis: *The analysis of a system, process, or activity to determine which parts of it are classified as real-value-added (RVA), business-value-added (BVA), or no-value-added (NVA).*

Value proposition: *A value proposition is a document that defines the benefits that will result from the implementation of a change or the use of an output as viewed by one or more of the organization's stakeholders. A value proposition can apply to an entire organization, parts thereof, or customers, or products, or services, or internal processes.*

Venture capitalist: Secures funding for innovation, evaluates and selects projects to receive resources, and guides implementation.

Vision: *A documented or mental description or picture of a desired future state of an organization, product, process, team, a key business driver, activity, or individual.*

Vision statements: *A group of words that paints a clear picture of the desired business environment at a specific time in the future. Short-term vision statements usually are between three to five years. Long-term vision statements usually are from 10 to 25 years. A visions statement should not exceed four sentences.*

Zones of conflict: Refers to the temporal zone and the operating zone of the problem—loosely the time and space in which the problem occurs.

Index

Page numbers followed by f, t, and n indicate figures, tables, and notes, respectively.